EXPLORING OPPORTUNITIES IN GREEN CHEMISTRY AND ENGINEERING EDUCATION

A Workshop Summary to the Chemical Sciences Roundtable

Paul Anastas, Frankie Wood-Black, Tina Masciangioli,
Ericka McGowan, and Laura Ruth, Editors

Chemical Sciences Roundtable
Board on Chemical Sciences and Technology
Division on Earth and Life Studies

NATIONAL RESEARCH COUNCIL
OF THE NATIONAL ACADEMIES

THE NATIONAL ACADEMIES PRESS
Washington, D.C.
www.nap.edu

THE NATIONAL ACADEMIES PRESS 500 Fifth Street, N.W. Washington, DC 20001

NOTICE: The project that is the subject of this report was approved by the Governing Board of the National Research Council, whose members are drawn from the councils of the National Academy of Sciences, the National Academy of Engineering, and the Institute of Medicine. The members of the committee responsible for the report were chosen for their special competences and with regard for appropriate balance.

This study was supported by the U.S. Department of Energy under Grant DE-AT01-94ER155535, the National Institutes of Health under Grant DHHS N01-OD-4-2139 (Task Order 25), and the National Science Foundation under Grant CHE-0621582.

Any opinions, findings, conclusions, or recommendations expressed in this publication are those of the authors and do not necessarily reflect the views of the organizations or agencies that provided support for the project.

International Standard Book Number-13: 978-0-309-10352-7
International Standard Book Number-10 0-309-10352-5

Additional copies of this report are available from National Academies Press, 500 Fifth Street, N.W., Lockbox 285, Washington, DC 20055; (800) 624-6242 or (202) 334-3313 (in the Washington metropolitan area); Internet, http://www.nap.edu.

Printed in the United States of America

THE NATIONAL ACADEMIES

Advisers to the Nation on Science, Engineering, and Medicine

The **National Academy of Sciences** is a private, nonprofit, self-perpetuating society of distinguished scholars engaged in scientific and engineering research, dedicated to the furtherance of science and technology and to their use for the general welfare. Upon the authority of the charter granted to it by the Congress in 1863, the Academy has a mandate that requires it to advise the federal government on scientific and technical matters. Dr. Ralph J. Cicerone is president of the National Academy of Sciences.

The **National Academy of Engineering** was established in 1964, under the charter of the National Academy of Sciences, as a parallel organization of outstanding engineers. It is autonomous in its administration and in the selection of its members, sharing with the National Academy of Sciences the responsibility for advising the federal government. The National Academy of Engineering also sponsors engineering programs aimed at meeting national needs, encourages education and research, and recognizes the superior achievements of engineers. Dr. Wm. A. Wulf is president of the National Academy of Engineering.

The **Institute of Medicine** was established in 1970 by the National Academy of Sciences to secure the services of eminent members of appropriate professions in the examination of policy matters pertaining to the health of the public. The Institute acts under the responsibility given to the National Academy of Sciences by its congressional charter to be an adviser to the federal government and, upon its own initiative, to identify issues of medical care, research, and education. Dr. Harvey V. Fineberg is president of the Institute of Medicine.

The **National Research Council** was organized by the National Academy of Sciences in 1916 to associate the broad community of science and technology with the Academy's purposes of furthering knowledge and advising the federal government. Functioning in accordance with general policies determined by the Academy, the Council has become the principal operating agency of both the National Academy of Sciences and the National Academy of Engineering in providing services to the government, the public, and the scientific and engineering communities. The Council is administered jointly by both Academies and the Institute of Medicine. Dr. Ralph J. Cicerone and Dr. Wm. A. Wulf are chair and vice chair, respectively, of the National Research Council

www.national-academies.org

CHEMICAL SCIENCES ROUNDTABLE

Cochairs

F. FLEMING CRIM (NAS), University of Wisconsin, Madison
MARY L. MANDICH, Bell Laboratories, Murray Hill, NJ

Members

PAUL ANASTAS, Green Chemistry Institute, Washington, DC
PATRICIA A. BAISDEN, Lawrence Livermore National Laboratory, Livermore, CA
MICHAEL R. BERMAN, Air Force Office of Scientific Research, Arlington, VA
APURBA BHATTACHARYA, Texas A&M, Kingsville, TX
LEONARD J. BUCKLEY, Defense Advanced Research Projects Agency, Arlington, VA
WILLIAM F. CARROLL, JR., Occidental Chemical Corporation, Dallas, TX
CHARLES P. CASEY (NAS), University of Wisconsin, Madison
JOHN C. CHEN, Lehigh University, Bethlehem, PA
ARTHUR B. ELLIS, National Science Foundation, Arlington, VA
GARY J. FOLEY, U. S. Environmental Protection Agency, Research Triangle Park, NC
TERESA FRYBERGER, Office of Science and Technology Policy, Washington, DC
ALEX HARRIS, Brookhaven National Laboratory, Upton, NY
SHARON HAYNIE, E. I. du Pont de Nemours & Company, Wilmington, DE
NED D. HEINDEL, Lehigh University, Bethlehem, PA
CAROL J. HENRY, American Chemistry Council, Arlington, VA
PAUL F. MCKENZIE, Bristol-Myers Squibb Company, New Brunswick, NJ
GEOFFREY PRENTICE, National Science Foundation, Arlington, VA
MARQUITA M. QUALLS, GlaxoSmithKline, Collegeville, PA
DOUGLAS RAY, Pacific Northwest National Laboratory, Richland, WA
GERALDINE L. RICHMOND, University of Oregon, Eugene
MICHAEL E. ROGERS, National Institutes of Health, Bethesda, MD
ERIC ROLFING, U.S. Department of Energy, Washington, DC
FRANKIE WOOD-BLACK, Conoco-Phillips, Ponca City, OK

National Research Council Staff

DOROTHY ZOLANDZ, Director
TINA M. MASCIANGIOLI, Program Officer
ERICKA M. MCGOWAN, Associate Program Officer
SYBIL A. PAIGE, Administrative Associate
DAVID C. RASMUSSEN, Senior Project Assistant

BOARD ON CHEMICAL SCIENCES AND TECHNOLOGY

Preface

Awareness of issues related to the environment—the need to conserve, the need for pollution minimization, the need to design for the future—have become part of the social dialog. It is seen in advertising: "green" in car commercials. It is seen at the grocery store: "paper or plastic?" It is seen in our personal energy use: "Do you choose the company that gets part of its electricity from renewable sources or standard resources?" It is part of the voting platforms—balancing the needs of having national parks with exploration and utilization of resources. Although these discussions are occurring in many different sectors of society, contradictory actions are also taking place. Most people still drive to work—increasing the need for more energy sources that are transportable. There is still a level of consumerism that leads to new waste streams, such as electronic waste (e.g., dead computers, cell phones that are no longer in vogue, personal data assistants). The list of such examples is long. This is not just an issue in the United States. Similar trends are occurring in Europe, Asia, and other parts of the world as we all strive for better standards of living without always considering the potential environmental impacts. All of these factors are drivers for the discussion of green chemistry and engineering. We need to understand the consequences of our actions, what the choices are, how the selection of one choice over another impacts our future, and how to develop and invent alternatives and solutions that improve the current state of our world.

In an effort to advance the discussion of green chemistry and engineering, the National Academies' Chemical Sciences Roundtable (CSR) held a workshop in November 2005 that was designed to look at the current state of green chemistry and green engineering education; to raise awareness about the tools that are available but may not yet be fully implemented across educational institutions; and to highlight promising new areas that are yet to be fully explored. This workshop was a chance to gather information, share ideas, and develop a platform from which the scientific and engineering community can address some particularly challenging issues.

This document summarizes the presentations and discussions that took place at the workshop. In accordance with the policies of the CSR, the workshop did not attempt to establish any consensus conclusions or recommendations about the needs and future directions to be taken, focusing instead on the issues identified by the speakers.

Understanding and knowledge are essential to developing a sustainable future. The chemical sciences and engineering community have a very special role to play in fulfilling that future by the development of new materials, understanding the toxicity of materials, developing new fuel sources, and understanding how chemical processes impact the environment. Yet, we are caught in between the present and implementing change in the future. Challenges are coming toward us at an ever faster pace, and it will take the energy, drive, and mental capacity of us all to meet them.

Paul Anastas and Frankie Wood-Black
Workshop Organizers

Acknowledgment of Reviewers

This workshop summary has been reviewed in draft form by persons chosen for their diverse perspectives and technical expertise in accordance with procedures approved by the National Research Council's Report Review Committee. The purpose of this independent review is to provide candid and critical comments that will assist the institution in making its published workshop summary as sound as possible and to ensure that the summary meets institutional standards of objectivity, evidence, and responsiveness to the workshop charge. The review comments and draft manuscript remain confidential to protect the integrity of the deliberative process. We wish to thank the following individuals for their review of this workshop summary:

Dr. Martin Abraham, University of Toledo
Dr. Joseph Fortunak, Howard University
Dr. Patricia Hogan, Suffolk University
Dr. Phillip Jessop, Queens University

Although the reviewers listed above have provided many constructive comments and suggestions, they did not see the final draft of the workshop summary before its release. The review of this workshop summary was overseen by **Dr. Jeffrey Siirola** of Eastman Chemical Company. Appointed by the Division on Earth and Life Studies, he was responsible for making certain that an independent examination of this workshop summary was carried out in accordance with institutional procedures and that all review comments were carefully considered. Responsibility for the final content of this workshop summary rests entirely with the authors and the institution.

Contents

1

Overview

A hot new topic in both chemistry and chemical engineering is green. Green chemistry is the design of chemical products and processes that reduce or eliminate the use and generation of hazardous substances.[1] Green engineering is the development and commercialization of industrial processes that are economically feasible and reduce the risk to human health and the environment. At the forefront of the green chemistry and engineering movement is Dr. Paul Anastas, director of the American Chemical Society (ACS) Green Chemistry Institute (GCI). According to the GCI, the overall goal of green chemistry and green engineering is to unleash "the creativity and innovation of our scientists and engineers in designing and discovering the next generation of chemicals and materials so that the chemicals and materials provide increased performance and value while meeting all goals to protect and enhance human health and the environment."

In this workshop, widespread implementation of green chemistry into undergraduate and graduate education was explored.[2] This workshop focused on the integration of green chemistry and engineering into the established and developing chemistry and chemical engineering curricula. Leading educators and industry managers showcased exemplary programs and provided a forum for discussion and critical thinking about the development, evaluation, and dissemination of promising educational activities in green chemistry. Speakers at the workshop:

- Provided an overview and current status of green chemistry education. They addressed how green chemistry and engineering bring value to the chemistry curriculum and why some educators in other disciplines choose to incorporate green chemistry and engineering educational principles into their teaching.
- Highlighted the most effective green chemistry educational practices to date, including government-industry collaborations and assessment activities in green chemistry.
- Discussed the most promising educational materials and software tools in green chemistry and engineering, including compelling industry examples that can be used as green chemistry and engineering teaching tools.

This summary is a compilation of the three main speaker sessions and the six breakout session discussions that allowed the participants to explore how to make green chemistry and engineering an integral part of curricula at all educational levels. The three main speaker session topics were (1) Current status; (2) Tools and materials; and (3) Where do we go from here?

The topics of the six breakout session discussions were:

1. Green chemistry and green engineering in future curricula;
2. What materials, programs, and tools are needed?
3. What is needed to achieve interdisciplinary approaches?
4. Green chemistry and green engineering industry and education;
5. Green chemistry and green engineering and the new faculty; and
6. Creating incentives, removing impediments.

The overall purpose of this summary is to be a resource for any educator who is interested in green science and technology education.

---

[1] ACS Green Chemistry Institute. Available at http://www.chemistry.org/portal/a/c/s/1/acsdisplay.html?DOC=greenchemistryinstitute\index.html.

[2] The views and opinions expressed in this the Green Chemistry and Engineering Education workshop and this workshop summary is not representative of the view of the Chemical Sciences Roundtable.

SETTING THE WORKSHOP STAGE: PRE-WORKSHOP PARTICIPANT SURVEY

As a precursor to the workshop, Dr. Anastas captured constructive ideas on how to address green education issues through an informal 10-question pre-workshop survey[3] of the workshop participants. Forty-three of the workshop participants—people from academe, industry, government, and nonprofit organizations—answered a mix of multiple-choice, yes-no, and open-ended questions. The questions covered many topics in green education, including who was interested, how it should be taught, who would benefit, and what mechanisms existed for funding. According to the survey results, in addition to helping teach technical issues, the main benefits of teaching green chemistry and green engineering were enthusiasm, continued interest, and increased job opportunities. The majority of participants also felt that integrating green chemistry and engineering throughout the four years of an undergraduate curriculum, is a more effective method for teaching green chemistry and engineering than having a single undergraduate course or waiting until the graduate level. In addition to the basic issue of funding mechanisms, other barriers for teaching green chemistry and engineering identified by the respondents included lack of tools and resources, already crowded curricula, and collegial resistance. The results of the pre-workshop survey were used by the workshop leaders to guide the discussions of what is being done at all levels of education and what can be done in the future to further green chemistry and green engineering education.

OPENING REMARKS

Workshop organizers Anastas and Wood-Black warmly welcomed the 75 attendees to the two-day discussion of green chemistry and engineering education. They explained the purpose and organization of the workshop.

Anastas explained that the time is right for leaders in green chemistry and engineering to push green concepts because the ideas of green chemistry and engineering are slowly being accepted within the broader scientific community. One example of the emerging interest in green approaches cited was the awarding of the 2005 Nobel Prize in Chemistry to Robert Grubbs, Richard Schrock, and Yves Chauvin "for the development of the metathesis method in organic synthesis" provided an excellent example of green chemistry and engineering. A second example he gave was the movement of the Green Chemistry Research and Development Act through both the U.S. House and Senate after passing the first hurdle of the House in April 2004.[4] A third example provided by Anastas was the placement of green chemistry education on the Carnegie Groups' agenda (e.g., Center for Sustainable Engineering).[5]

Anastas closed his remarks by discussing impediments to innovation. He explained that change can come much more slowly than anyone would expect because people do not like to do things differently from the way they have done them before. New ideas and new perspectives often face harsh opposition. He led the audience in considering some amusing historical examples of mistakes made by a few of our greatest scientific leaders:

- Lord Kelvin, discoverer of the temperature scale named for him, denied his date for the age of the earth (24 million years old) was wrong even after radioisotope dating had demonstrated his value to be false;
- Mendeleev, inventor of the periodic table, denied the existence of radiation and the electron; and
- J. J. Thompson, discoverer of the electron, adhered to the belief in the existence of the "ether," which "is as essential to our lives as the air we breathe," long after this concept was disproved.

[3]A list of the 10 questions and tabulated answers are listed in Appendix A.

[4]Green Chemistry Research and Development Act of 2005. Available at http://thomas.loc.gov/cgi-bin/query/z?c109:h.r.1215.

[5]http://www.csengin.org/.

2

Current Status

In this session three main speakers and a panel of additional speakers were asked to provide an overview of the current status of green chemistry and engineering education by addressing how green chemistry and engineering bring value to the chemistry and chemical engineering curricula and to consider why some educators choose to incorporate or not incorporate green chemistry and engineering educational principles into their teachings.

MAIN SPEAKERS

The first speaker, Dr. David Allen (director, Center for Energy and Environmental Resources, University of Texas, Austin), gave a presentation titled "Green Engineering: Environmentally Conscious Design." He described the framework used at his center as an example of the current status of green engineering. This framework incorporates green concepts into chemical engineering and other initiatives to reformulate the engineering curriculum.

According to Allen, the evolution of green engineering began 20 years ago when the chemical engineering community began exploring waste minimization. In the late 1980s and early 1990s there was a considerable amount of commitment to bringing the concepts of waste reduction into the design of chemical processes and chemical products. The idea of waste reduction eventually evolved into pollution prevention. In the mid-1990s a series of textbooks and course modules on pollution prevention began appearing. In 2000 the U.S. Environmental Protection Agency, Allen, and some of his colleagues established a partnership to develop green engineering materials specifically for the chemical engineering curriculum. Allen stated that the current and future education focus should progress from greening the chemical engineering curriculum to incorporating some green concepts into other engineering disciplines.

Allen went on to identify two tools he uses when teaching green engineering: (1) assessment and (2) improvement. He uses assessment tools to determine what constitutes a green product or process and improvement tools to answer the questions, "Will new engineering design tools be necessary, or will our existing tools that allow us to minimize mass and energy consumption be sufficient?"[1] Allen said that it is possible to apply assessment to a variety of design stages and scales (i.e., molecular, process, and system scales), but that determining whether a process or product is green through assessment is not as simple as it might seem. The potential environmental impacts are considered when completing an assessment of a particular chemical process or product. However, comparing one product or process with another is difficult because most products and processes have unique fingerprints.

To emphasize the complexity of making such assessments, Allen provided the audience with a typical chemical engineering problem given to undergraduate students: "You have a vent stream that contains, in this case, two compounds, say toluene and ethyl acetate. You don't want to emit this to the atmosphere. So, you are going to use an absorbing column. That absorbing column contacts your gas vent stream with absorbing oil, captures those emissions, or at least some fraction of those emissions. Then you would send the material that has been absorbed in this absorbing column to a distillation column. You recover the materials that you have absorbed, and you recycle the oil back to the absorption column, a very simple chemical engineering process, junior level material." According to Allen, the problem

[1]Allen, D., and D. Shonnard. 2001. Green engineering: Environmentally conscious design of chemical processes and products. *AICHE Journal* 47(9):1906-1910.

with this approach to capturing emissions from the chemical process is that a large amount of energy is expended. It is possible that there is another process that does not expend as much energy, but it may have some other adverse effect. Carrying out an assessment of a chemical process or product may give an ambiguous result such as in the example provided, but at the very least an assessment can help identify the potential limitations of the process. Allen said that he also provides his students with screening metrics to complete an assessment of such items as environmental impacts, costs, and sustainability metrics.

Deciding where improvements for products or processes can be made requires the consideration of whether new engineering design tools are necessary or whether existing tools that allow us to minimize mass and energy consumption are sufficient. According to Allen, most improvement for traditional systems is achieved through the use of conventional tools of process design, but the examination of new systems will require the development of new tools for improvement. Some new tools of improvement for integrating material and energy flows across industrial sectors include sustainable technologies, mass-energy balances, life-cycle assessments, and national scale material and energy flows.

In closing, Allen highlighted some specific tools "designed to dovetail with the fundamental reform that is occurring in chemical engineering education." These tools should be actively disseminated throughout the scientific community. He said that the Massachusetts Institute of Technology is leading the advancement of undergraduate chemical engineering curriculum[2] through the discipline-wide initiative Frontiers in Chemical Engineering Education. According to Allen, the initiative is exploring the extension of several basic themes in collaboration with other branches of engineering and other audiences: (1) the focus of chemical engineers in the future, (2) multiscale engineering, (3) molecular transformations, and (4) sustainable systems engineering.

The second speaker in this session was Dr. James Hutchison, professor of chemistry and director of the Materials Science Institute at the University of Oregon, who described his green organic chemistry laboratory course. His presentation was titled "Green Chemistry Education Status: Lessons from the Organic Chemistry Laboratory Experience." Hutchison explained that his goal at his institution is to accomplish "broad implementation of green chemistry in the curriculum both at the undergraduate and graduate level," and his course is just one step toward achieving this goal. Over the course of teaching this laboratory series, Hutchison developed a student laboratory manual, "Green Organic Chemistry: Strategies, Tools, and Laboratory Experiments."[3] Using this manual, students perform green chemistry experiments and learn 19 concepts. Topics in the manual include:

- Identification of chemical hazards;
- Chemical exposure and environmental contamination;
- Evaluation of chemical hazards;
- Introduction to green chemistry;
- Alternative solvents;
- Alternative reagents;
- Reaction design and efficiency; and
- Alternative feedstocks and products.

For example, in the development of the experiments to find greener alternatives, Hutchison includes molecular assessment to observe potential hazards or inefficiencies and to find and test alternatives. Hutchison has found that this process teaches students how to develop greener laboratory experiments while performing them (see Figure 2.1).

Hutchison identified several challenges in implementing green chemistry in an already crowded curriculum. Three of the challenges are: (1) developing new experiments that illustrate green chemistry concepts and are effective in teaching labs; (2) developing state-of-art concepts that also integrate essential chemistry concepts with green chemistry; and (3) providing a flexible option for integrating green chemistry into the existing curricular framework. In an effort to address these challenges Hutchison suggested that the quality of teaching be ensured by thorough testing, a wide range of choices in the curricular framework, and replacing old material with new material.

Integrating green chemistry into the organic laboratory at the University of Oregon revealed several incentives for implementing the greener alternatives. First, the amount of waste generated from experiments has significantly decreased. Second, university and community public relations are improved. The University of Oregon's green chemistry program has generated 25 globally published journal articles. The green chemistry program has also enhanced student recruiting at both the undergraduate and graduate levels. Third, the classes were an opportunity to upgrade curricula and facilities. Because the green experiments do not require fume hoods, the laboratory atmosphere can be designed to be more inviting to students and provide a better view of the entire laboratory environment. Such improvements in the teaching environment are particularly attractive to a school with older facilities (e.g., a community college with a 40-year-old laboratory that may have inadequate ventilation). Fourth, increased safety, decreased liability, and reduced energy costs are all major incentives to implementing green chemistry into a curriculum.

The final main speaker in this session was Dr. Steven Howdle, the chair of chemistry at the School of Chemistry at the University of Nottingham. Howdle discussed the divide between chemistry and chemical engineering in his presentation titled "Mind the Gap: Bridging the Divide Between Chemistry and Engineering." Howdle explained how he developed the Green Chemistry for Process Engineering program as a new undergraduate degree at the University of Nottingham. The program has been running for four years. The program brings modules from chemistry and chemical engineering together to train

---

[2]http://mit.edu/che-curriculum.

[3]Doxsee, K., and J. Hutchinson. 2004. *Green Organic Chemistry: Strategies, Tools, and Laboratory Experiments*. 1st ed. Florence, KY: Brooks/Cole.

undergraduates in aspects of chemistry and chemical engineering. According to Howdle, the first year has a module that presents "hot" green topics and serves as a way to explain why the classes are important beyond the classroom to any student, not just chemists. In addition to chemists, students from other majors (e.g., music, English) are taking the module, which is now the most popular module on the campus at Nottingham. The chemical engineers, however, cannot fit the class into their densely packed program.

Although his course is very popular now, Howdle pointed out that the overall program has not been overwhelmingly successful in that "only two students per year for the last four years have signed up full-time for the course." Despite this unpleasant result, however, he said that other universities are following the example of this program by developing courses to bring chemistry and chemical engineering together.

PANEL SPEAKERS

While most of the speakers in this session were experienced professors and career professionals, Dr. Amy Cannon, a recent graduate of the Green Chemistry Ph.D. Program at the University of Massachusetts (UMASS), provided a different point of view. Cannon is the first graduate of the "the world's first green chemistry Ph.D. program" at UMASS Boston. This program was started in 2001 and currently has 15 students. In addition to core chemistry courses, the program requires courses in toxicology and risk assessment, environmental fate and transport, environmental economics, and environmentally benign synthesis. In addition, students are required to defend three independent research proposals to a committee.

Cannon discussed her experience entering the workforce as a new graduate in green chemistry. She is employed by Rohm and Haas's Electronic Materials Division and designs waveguide materials for optical electronic devices. Cannon also teaches the Introduction to Green Chemistry course at UMASS Lowell and an undergraduate and online course at UMASS Boston.

Dr. Berkeley Cue, a retired pharmaceutical executive and Green Chemistry Governing Board member, was able to provide another dimension to the current status of green chemistry education. In his talk titled "What Industry Can Do to Encourage Green Chemistry Education: A Pfizer Case Study" Cue indicated that industry is interested in promoting green chemistry because industry now recognizes its social responsibility to the community.[4] Cue described Pfizer's development of the Pfizer Groton Labs Green Chemistry Workshop. In the workshop, 25 to 30 students, both undergraduates and graduates, are invited to the Groton Labs where they are introduced to the pharmaceutical industry and learn how pharmaceutical research and development is performed.

Pfizer also has a few programs targeted at middle school students. Green Chemistry and Environmental Sustainability provides a 10-day module that contains exercises, readings, as well as experiments in science, math, language and arts, and social studies. The program has been mapped to national education standards. There is currently a 10-school pilot program in southeast Connecticut, and Pfizer expects a national rollout near Pfizer research sites in 2006. Samjam, a science and math jamboree, and Smart Science and Math are two more programs for middle school students sponsored by Pfizer. More than 3,000 students a year participate in the Samjam modules, and more than 200 Pfizer employees take time out to produce and run experiments for middle school students.

Cue highlighted other current green chemistry efforts, such as the elementary school-level coloring book "Pollution Solution: A Green Chemistry Story." The coloring book was developed by a group of organic chemistry students at Suffolk University and was based on SEA-NINE 211™, a compound that received the 1996 Green Chemistry Award.[5] Other notable green chemistry efforts are the ACS Green Chemistry Summer School program at McGill University and Pfizer's internal award recognition program.

Cue closed with an action item for industry: "In every job advertisement for chemists and chemical engineers, add one sentence: A knowledge of green chemistry (or green engineering) is desirable. If the students respond to our challenge to learn green chemistry, industry has to respond by hiring them."

Dr. Kenneth Doxsee (National Science Foundation and the University of Oregon) discussed the current existence of green chemistry education in educational institutions. Doxsee highlighted "green islands," which are "relatively small pockets of activity in green chemistry education." These islands are Carnegie Mellon University, Gordon College, Hendrix College, University of Massachusetts, University of Oregon, University of Pittsburgh, and University of Scranton. Doxsee indicated that the connections between these islands are very important, but it is even more important to expand green chemistry into more research extensive universities 1 (R1).[6]

Doxsee described how the University of Oregon hosts a Green Chemistry Education Workshop[7] that focuses on implementing green chemistry into organic chemistry cur-

[4]Rottas, M., M. Kirchoff, and K. Parent. 2004. Pfizer works with future scientists to promote environmentally responsible science. *inChemistry Magazine*. 13(4):17.

[5]Rohm and Haas was recognized for its development of SEA-NINE®211 antifouling agent, an effective and more environmentally acceptable ingredient for use in marine antifouling paints, compared with many currently used biocides.

[6]The term "R1" is used in the United States to describe Research Extensive Universities 1.'R1s offer a full range of baccalaureate programs with research having a high priority. There are currently 88 public and private universities classified as R1s.

[7]http://chemistry.gsu.edu/CWCS/green.php.

riculum. At the University of Oregon workshop, faculty members try new experiments, learn approaches to incorporate green chemistry into their curriculum, and network with other educators. This workshop is jointly sponsored by the University of Oregon, National Science Foundation (NSF), and the NSF-sponsored Center for Workshops in the Chemical Sciences. According to Doxsee, the University of Oregon has been hosting this workshop for five years with the sixth year in summer 2006. He said that there is a tremendous amount of interest from community colleges, high schools, and four-year teaching colleges, but the workshop lacks representation from R1 institutions, the top funded major research institutions in the country.

Doxsee shared his interest in getting the R1 institutions to buy into green chemistry for several reasons. First, acceptance by R1 institutions may increase acceptance in the broader education community. Second, major institutions train a large number of students. Third, they provide a considerable amount of intellectual capital to major industrial employers. Fourth, R1 schools are training the next generation of faculty. According to Doxsee, the lack of attendees from R1 institutions at the organic chemistry laboratory workshops is due to their attitude toward green chemistry. He believes that there is a reluctance to move away from the traditional method of teaching at R1 institutions. Doxsee also believes R1 institutions may feel they do not need any help with green chemistry implementation and concepts, they are just not interested in green chemistry, or think that green chemistry is a bad idea.

Despite the reluctance at many R1 institutions, Doxsee pointed out signs of hope in gaining support from some R1 institutions. The support includes representation of R1 schools, such as MIT and Cornell, at this workshop; research endeavors in graduate programs at research intensive university graduate programs; and international workshops that provide a platform to introduce new educational materials to educators where high levels of R1 representation are common. Doxsee pointed out that although these endeavors are positive, because of their rarity, they do not make as much of an impact.

In addition to highlighting the University of Oregon's organic chemistry laboratory and the supplementary laboratory manual, Doxsee mentioned a German-authored textbook that will also be published in English, titled *Chemistry Experimentation for All Ages*.[8] The textbook focuses heavily on microscale chemistry and has at least one chapter that discusses green chemistry. The book targets students at elementary levels, including kindergarten, through high school. In advance of publication the German editor has already introduced the book to high school students in Germany.

[8]Schwarz, P., M. Hugerat, and M. Livneh. 2006. *Chemistry Experimentation for All Ages*. Arab Academic College for Education in Israel: Haifa, Israel.

In closing, Doxsee emphasized that green "educational needs go beyond our undergraduates and beyond the K-12 level. We need to educate industry; we need to educate our colleagues."

The final panel speaker of this session, Dr. Tyler McQuade, from Cornell University, described a different method of green education. He has a program that encourages postgraduates to focus on the business side of green chemistry and engineering with the goal of developing and educating green entrepreneurs and innovators. His group at Cornell, which is a combination of chemistry, biology, and materials science engineering, works on innovations in industry using the field of green chemistry. McQuade highlighted the many different topics his group covers, which include:

- Commerce issues;
- Patenting;
- Interactions with industry;
- Business idea competition;
- Interactions with business schools;
- Interaction with campus entrepreneur organizations; and
- Reaction efficiency with technologies, such as telescoping.

BREAKOUT SESSIONS

On the second day of the workshop, planned breakout sessions began that allowed participants to delve deeper into the issues surrounding green chemistry and engineering. Workshop participants were pre-assigned to breakout groups and the results of those breakout sessions that correspond with the current status of green chemistry and engineering education are listed below.

Green Chemistry and Green Engineering and the New Faculty

During this breakout session, participants discussed faculty efforts to implement green principles. Participants felt that existing faculty members view new faculty who bring green concepts into the curriculum either favorably or with ambivalence. The ambivalence stems from concerns about the rigor of research despite the use of green principles. Because the new faculty's green efforts are commonly not recognized one way or the other, those who do try to incorporate green principles are not sure what type of impact they are making on the department. On the other hand, green principles are seen as a positive addition in cases where new students are attracted to the institution or a school is recognized due to green chemistry or engineering.

The breakout group participants also discussed the impact of teaching green principles on the tenure process. Some believed teaching or incorporating green chemistry and engineering into the curricula helps graduates in their future careers and can also help in acquiring research funding.

A summary of key roadblocks for new or tenured faculty trying to adopt green chemistry and engineering include traditionalists, lack of guidance or mission statement from professional society, lack of funding, and lack of publication in top journals. Addressing these roadblocks, collaborating with green chemists and engineers at other institutions, and developing a Green Chemistry Institute workshop for new faculty may provide inspiration and therefore encourage new faculty to incorporate green chemistry and engineering concepts into their curricula.

Green Chemistry and Green Engineering Industry and Education

Industry views green chemistry and engineering in different ways. Green thinking could potentially be a successful business investment. Creating a new product that can be sold at a higher price, because it has a more intricate development process that requires a higher level of expertise or can be marketed as being green, and decreases waste is favorable for the chemical industry's reputation and profit margin. Green thinking could also be added to the industry's current sustainable development efforts. On the other hand, green chemistry and engineering could lead to the development of new regulations or be seen as alternative forms of environmental chemistry or sanitary engineering, both of which some companies view as energy intensive efforts without many positive benefits.

Participants had varying answers to the question, "Are green chemistry and engineering practitioners readily finding employment?" Some participants believed that more green chemistry graduates would propel the industry to seek out this expertise. Some participants, however, believed that green chemistry and engineering practitioners are not finding employment because large companies can depend on smaller companies to provide green expertise on an ad hoc basis. The cost is probably much less than directly hiring green chemists or engineers as full-time staff because the company must provide a competitive salary and benefits. It is important to note that the definition of a green chemist or engineer is still a gray area; some scientists practice green chemistry or engineering but do not label themselves as green chemists or engineers.

Industry and academia are promoting green chemistry and engineering to make their respective organizations more competitive. Industry is greening R&D programs, while academia is developing green chemistry and engineering programs.

The participants identified the following actions that may aid in addressing issues related to green chemistry and engineering in industry and academia:

- The federal government and nonprofit organizations could promote green principles to the general public in two ways: (1) through entertainment and educational events, and (2) by teaching green chemistry and engineering to young children, to potentially influence the next generation to carry green chemistry and engineering into the future.
- Professional societies could provide more funding and create more interest through promotion, for example, at professional society meetings and conferences or through society-sponsored journals, to place more emphasis on green chemistry and engineering.

3

Tools and Materials

In the next portion of the workshop, speakers and panel members focused on effective green chemistry and engineering educational programs, materials, and teaching tools, including computer software. The session started with talks by four main speakers, followed by four panel speakers.

MAIN SPEAKERS

The first speaker, Dr. Julie Haack from the University of Oregon, provided the audience with her presentation titled "Community-Based Approach to Educational Materials Development." Haack explained that a community-based approach is "a community that really empowers people to participate" and should encourage increasing access to information and resources; enhancing the capabilities of the members through the exchange of knowledge and experience; and facilitating innovation.

Haack explored some examples of these community-based activities in her presentation. One example is Greener Education Materials for Chemists (GEMs),[1] a database of educational materials focused on green chemistry. This Internet-based database holds a searchable collection of green chemistry books, articles, demonstrations, courses, laboratory exercises, and other databases (see Figure 3.1). The GEMS database serves the function of increasing access to information and resources related to green chemistry, enhancing capabilities by providing quality materials, and decreasing the potential barriers to communication.

In addition to the GEMS database, Haack emphasized the importance of incorporating green chemistry through other means. The University of Oregon in collaboration with Worcester State College is in the early stages of developing a high school distance education program. The development comprises several parts: (1) modifying or coordinating existing materials; (2) designing new materials, (e.g. podcasts, games); (3) course design collaborative; and (4) information dissemination channels.

Another example Haack mentioned is the text *Chemistry for Changing Times*,[2] a chemistry textbook for nonchemistry majors. The nonchemistry major student population includes students in education, business, and health fields, such as physical therapy, art, and history. Typically these students are trying to satisfy a science requirement for the university's core requirements and will not take any additional chemistry. The textbook has very little math and focuses on concepts. The new edition has 10-12 new educational modules that cover green chemistry.

The establishment of the Ambassador Site Project is another example of the University of Oregon's efforts in green chemistry education. This project grew from University of Oregon's Green Chemistry and Education Workshop. At the workshop Haack and her colleagues observed that many faculty members had modified laboratories to remove environmental hazards but were not published as green alternatives. Unfortunately, faculty members were not sharing these laboratories with students or their colleagues. This prompted collaboration between Haack, her colleagues at Oregon, as well as others who were successful in incorporating green chemistry into their curriculum, such as Liz Gron and Tom Goodwin (Hendrix College), Margaret Kerr (Worcester State College), and Irvin Levy (Gordon College). Their collaboration resulted in the development of ambassador sites that utilize a community-based approach which,

[1]http://greenchem.uoregon.edu.

[2]Hill, J. W., and D. K. Kolb. 2003. *Chemistry for Changing Times*. Upper Saddle River, NJ: Prentice Hall. Available at http://wps.prenhall.com/esm_hillkolb_chemistry_10.

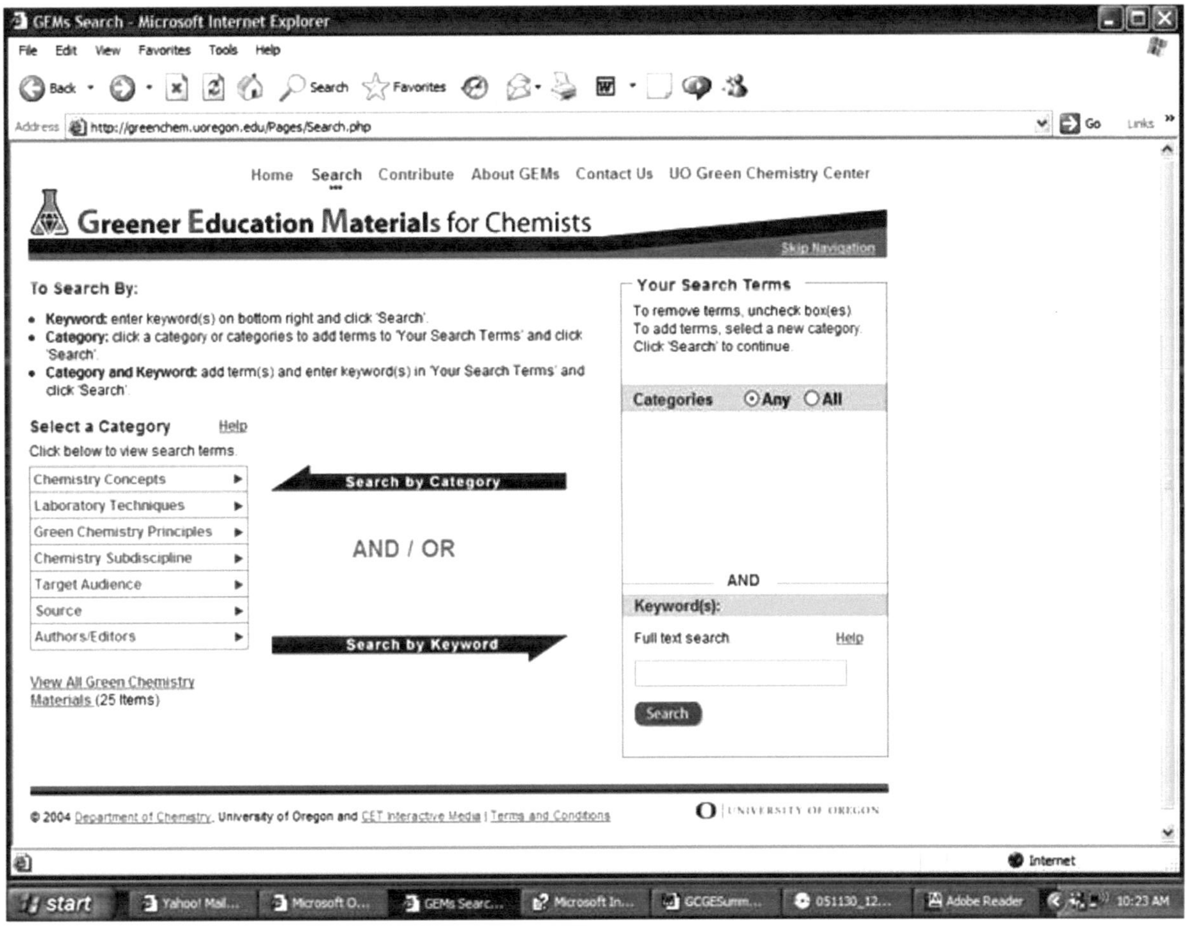

FIGURE 3.1. Example Web shot of searching the GEMS website. SOURCE: Haack, J. 2005. A Community-Based Approach to Educational Materials Development. Presentation at the National Academies Chemical Sciences Roundtable Green Chemistry and Engineering Education Workshop. November 7, 2005.

according to Haack, empowers people to participate at different levels to facilitate the incorporation of green chemistry materials into the curriculum, increases access to information and resources, and enhances the capability of the group through participation and provides a foundation or framework for innovation.

The educational ambassador sites will create new materials, write grants, offer mentoring and professional development, and distribute materials.

The next speaker in this session, Dr. David Shonnard (Michigan Technological University) began his talk by giving a definition of green engineering as "the design and commercialization and use of processes and products that are both feasible and economical, while minimizing risk to the environment and to human health and also the generation of pollution at the source." Shonnard discussed using the "box" concept, where inputs and outputs are balanced within the context of conservation laws to develop governing equations as a teaching tool (see Figure 3.2). One could complete analyses at differing scales or levels to yield useful information using the "box" concept.

In addition to the "box" concept, Shonnard discussed computer-aided assessment and improvement tools that can be used in green engineering. According to Shonnard, "computer-aided tools can help inform process or product design early on through estimation of chemical process and environmental properties, later through process simulation and environmental fate modeling, and ultimately by using process integration and multi-objective optimization." The tools can be used for a range of scales, including molecular, process, national, or global. Green Engineering incorporates these tools in a hierarchical design sequence (see Figure 3.3).

Some of the computer-aided tools that Shonnard highlighted in his talk included:

- Tools for early design assessment to predict environmental properties, investigate green chemistry alternatives, and design molecules with lower environmental impacts.

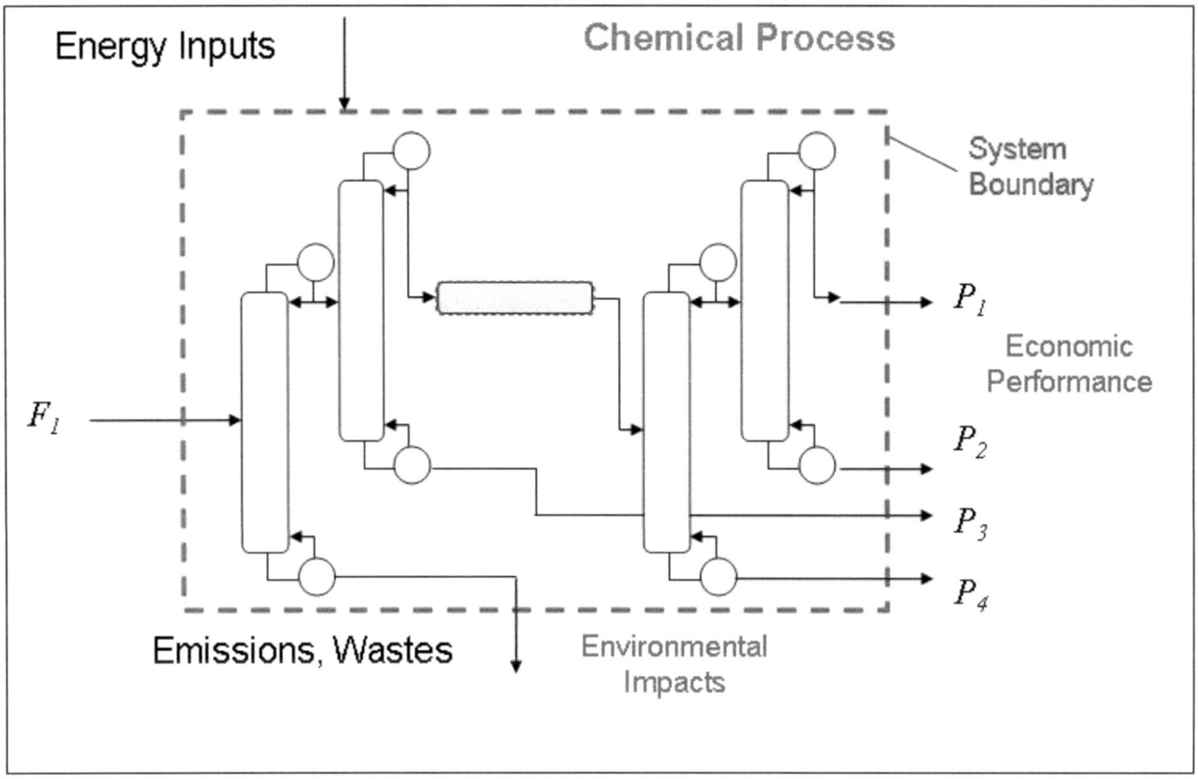
Energy Inputs
Chemical Process
System Boundary
F_1
P_1
Economic Performance
P_2
P_3
P_4
Emissions, Wastes
Environmental Impacts

A

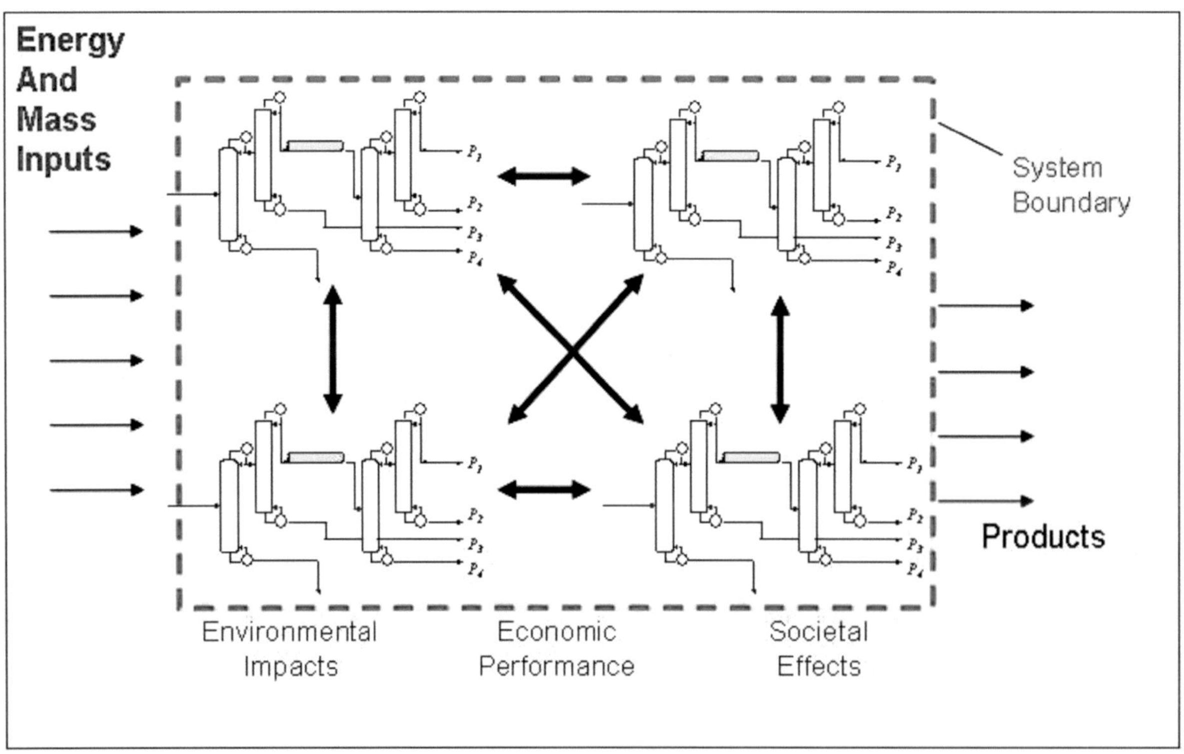
Energy And Mass Inputs
System Boundary
Products
Environmental Impacts
Economic Performance
Societal Effects

B

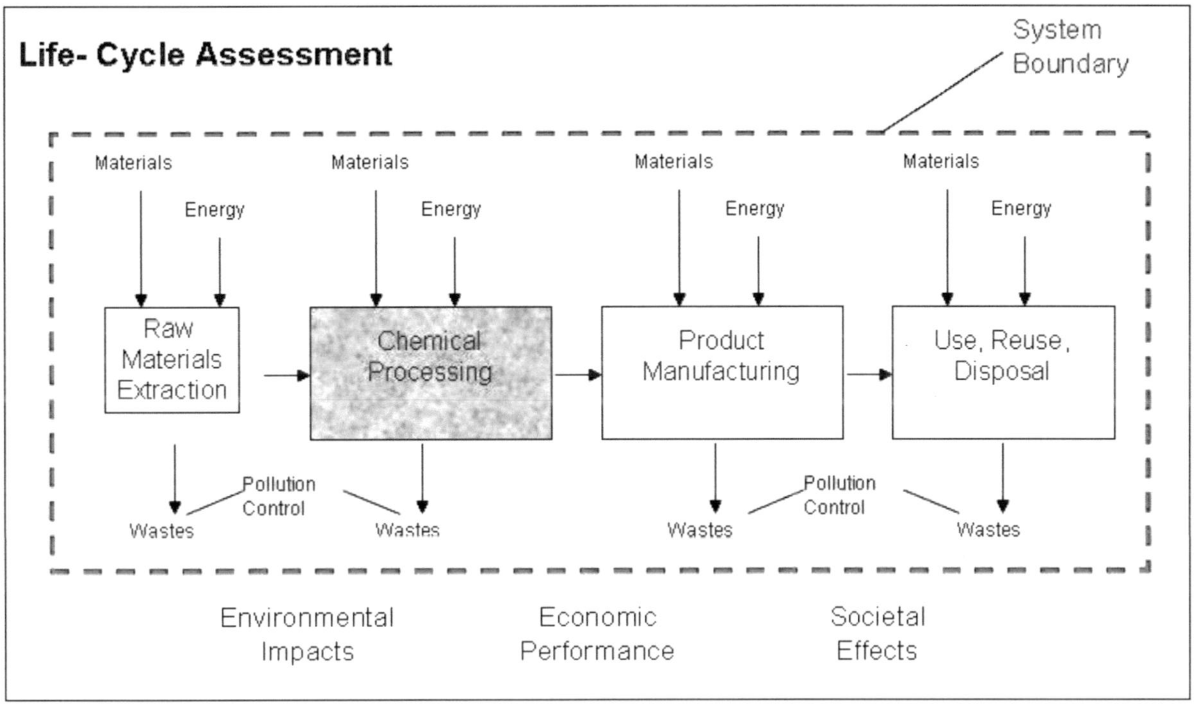

C

FIGURE 3.2 (A) Box concept at the macroscale, (B) Box concept: Exchanges within and between facilities, (C) Box concept: Beyond the plant boundary. SOURCE: Shonnard, D. 2005. Tools and Materials for Green Engineering and Green Chemistry Education. Presentation at the National Academies Chemical Sciences Roundtable Green Chemistry and Engineering Education Workshop. November 7, 2005.

➢ EPI Suite looks at physical and chemical properties and environmental fate estimation models developed by the EPA.[3]

➢ The Green Chemistry Expert System (GCES)[4] can also be used to design green chemistry reactions and reaction conditions.

➢ The Program for Assisting the Replacement of Industrial Solvents (PARIS II)[5] software has been created for the purpose of finding replacements for currently used solvents that have similar properties but are less harmful to the environment.

- Tools for environmental impact assessment of process designs.

➢ Simultaneous Comparison on Environmental and Non-Environmental Process Criteria (SCENE).[6]

➢ Waste Reduction Algorithm (WAR).[7]

➢ Tool for the Reduction and Assessment of Chemical and Other Environmental Impacts (TRACI).[8]

- Tools that aid in the estimation of pollutant release from processes to the air.

➢ Air CHIEF CD[9] for emission factors for major equipment plus fugitive sources.

➢ TANKS 4.0—program from EPA[10] for storage tanks.

➢ WATER8—on Air CHIEF CD[11] or EPIWIN for wastewater treatment.

➢ CHEMDAT8—on Air CHIEF CD for treatment storage and disposal facility (TSDF) processes.

Most of these software programs are available free of charge or for a very small fee.

Other educational materials Shonnard highlighted were a book and Web site. His book *Green Engineering: Environmentally Conscious Design of Chemical Processes*, which was developed in collaboration with David Allen, contains an aggregate of green engineering Web resources, software tools, and online databases. The Web site Shonnard de-

[3]http://www.epa.gov/oppt/exposure/docs/episuite.htm.
[4]http://www.epa.gov/oppt/greenengineering.
[5]http://www.epa.gov/nrmrl/std/mtb/paris.htm.
[6]http://www.aiche.org/sache/.
[7]http://www.epa.gov/oppt/greenengineering/software.html.
[8]http://www.epa.gov/ORD/NRMRL/std/sab/traci/.

[9]http://\t "_parent" www.epa.gov/ttn/chief/airchief.html.
[10]http://www.epa.gov/ttn/chief/tanks.html.
[11]http://\t "_parent" www.epa.gov/ttn/chief/airchief.html.

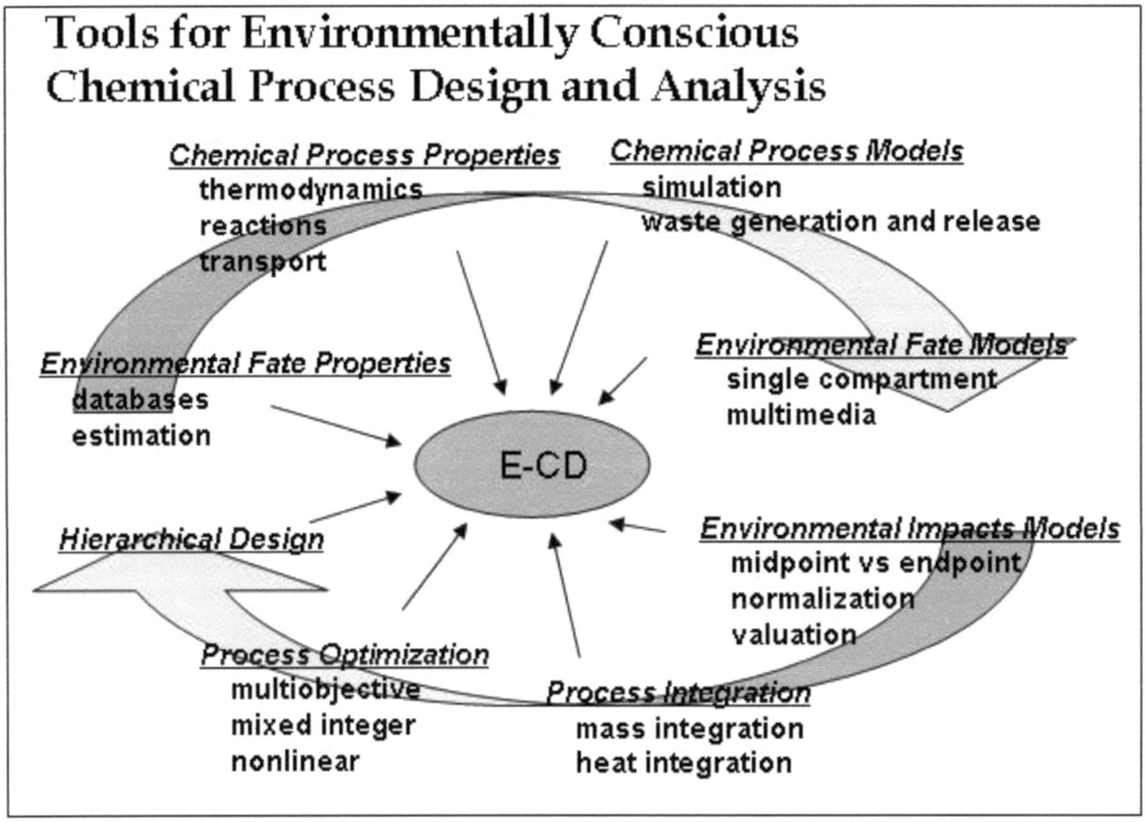

FIGURE 3.3 Schematic of David Shonnard's tools for environmentally conscious chemical process design and analysis. SOURCE: Shonnard, D. 2005. Tools and Materials for Green Engineering and Green Chemistry Education. Presentation at the National Academies Chemical Sciences Roundtable Green Chemistry and Engineering Education Workshop. November 7, 2005.

scribed was the *Green Engineering Website for Educators and Students* that was developed by Rowan University through the American Society for Engineering Education Green Engineering program. The Environmental Protection Agency and National Science Foundation provided funding for the site. This site contains a variety of resources: green engineering Web sites; announcements of green engineering journal publications, workshops, and presentations; links or references to related software; and courses or modules in green engineering for instructors. The undergraduate modules have been developed to aid instructors to integrate green engineering concepts into traditional engineering courses at all undergraduate levels.

The next speaker to discuss tools and materials for green chemistry and engineering education was Dr. John Andraos from York University. Andraos discussed his chemistry course, Industrial and Applied Green Chemistry, which is offered as an advanced course at the third-year level. Andraos stated, "I am one of the proponents who believe that it should be taught a little later so that students have acquired a real mastery of the subject." He explained that there are two prerequisites for the class: (1) second year organic chemistry with a minimum C grade plus brush-up quiz and (2) a science library resource workshop and quiz. The course is divided into seven sections:

1. Chemistry in society gives a historical account of chemistry by showing the connections between people and ideas;
2. Survey of modern concerns in which the students gain an accurate account of current issues in the industry by surveying scholarly literature;
3. Dyestuffs;
4. Green chemistry;
5. Pharmaceutical industry;
6. Industrial feedstocks; and
7. Chemistry of everyday experience.

The course has many components, such as Chemistry and Society, Development of Industrial Chemistry, and Genealogy, to connect chemistry to history, world events, and real-case problems. Students are required to research resources such as journal articles, society news magazines, books, and patent literature to enhance skills in decision making, interdisciplinary problem solving, quantitative reasoning and evaluation.

Andraos explained that he wants to encourage self-discovery through this independent learning process. In the business area, topics such as economic impacts, patents, and confidentiality agreements are reviewed as further examples of how chemistry is connected to society. The course also contains a career development component as well as alumni speakers. The coursework for the class comprises biweekly quizzes, four problem sets, one written assignment, one oral assignment, and one final exam. The written assignment is a rigorous critiquing of a synthesis or manufacturing of target product or process according to green criteria written in a journalistic style. The topic is the student's choice. Andraos commented that students come to the class thinking industry is the "bad boy" but go away with a more informed picture.

The final main speaker in this session was Dr. John Warner from the University of Massachusetts, Lowell, who is the founder of the "world's first green chemistry Ph.D. program." In his talk Warner discussed different aspects of the Ph.D. program and how his program teaches people how to do green chemistry. Warner recited Russian poetry as the introduction to his talk. After reciting three poems in Russian, he asked the question "Can we all be Russian poets since we have seen three examples?" He used this example to demonstrate that examples are useful but do not make us experts in a subject, green chemistry in particular.

Warner explained that although he feels compelled to teach green chemistry, when he was considering how to teach the subject he did not think that integrating green chemistry into existing curricula was the best mode of action. Therefore, he created a new, independent program in green chemistry that focuses on research to avoid obstacles in integrating green chemistry into existing curricula. His program is not located in the college of sciences, the college of engineering science, or the college of health and environment. Each college has representation on the Center for Green Chemistry board of advisers, but the center and its program stand alone.

In addition to research, the program Warner described consists of core and elective courses. The students are required to complete five core chemistry courses:

1. Introduction to Green Chemistry;
2. Mechanistic Toxicology;
3. Sustainable Materials Design;
4. Environmental Law and Policy; and
5. Experimental Conceptualization.

With the addition of electives and other required courses, a total of 12 classes are required. Students take five cumulative exams throughout the program, which are written by influential leaders in green chemistry from outside the program, such as Paul Anastas and Berkeley ("Buzz") Cue. An additional requirement in this program is that all students must defend three research proposals that must be orthogonal to their laboratory work. At this point students can opt to acquire a terminal master's degree or become doctoral degree candidates. If the latter is chosen, candidates immediately give a dissertation seminar describing their research to the entire university's research community. As stated by Warner, this path is chosen because too often in chemistry, we wait until the end of a student's academic career to find out what he or she has been doing for the last three or four years in the lab.

The options for research in the program are one of the seven areas in the Center for Green Chemistry:

1. Crystal engineering;
2. Noncovalent derivitization;
3. Photo polymers;
4. Ambient metal oxide semiconductors;
5. Reaction design;
6. Medicinal chemistry; or
7. Educational research.

One interesting aspect of the program, Warner noted, is the education research requirement for the program. All Ph.D. students must participate in community outreach at the K-12 level a minimum of once per month. The students receive no compensation or credit for this community outreach, but according to Warner, "It instills in them the sense that this is what people should do and when they leave, hopefully, whether they go into industry or academia this model follows with them and they see this is a requirement in their lives to be reaching out to the community."

PANEL SPEAKERS

The panel discussion on tools and materials for green chemistry and education began with Dr. Michael Cann from the University of Scranton. Cann presented tools and materials for infusing green chemistry into the undergraduate lecture curriculum. Cann believes there are three things needed to mainstream green chemistry: (1) insertion of green chemistry into mainstream chemistry courses; (2) faculty who teach these courses to develop modules on green chemistry related to topics already covered in their course; and (3) make it easy for other faculty to do the same by providing access to materials (e.g., place materials on the Web).

A starting point for Cann was the development of the book *Real World Cases in Green Chemistry*[12] with coauthor Marc Connelly. They designed the book to be used in a variety of ways. It contains descriptions of 10 projects that have won or been nominated for Presidential Green Chemistry Challenge awards. The book can also serve as a resource for

[12]Cann, M. C. and M. E. Connelly. 2000. *Real World Cases in Green Chemistry*. Washington, DC: American Chemical Society.

anyone wishing to be better informed about specific ways that the redesign of chemical products and processes is preventing pollution and solving environmental problems.

In his quest for mainstreaming green chemistry, Cann identified two of his objectives: to develop modules and to make green chemistry accessible to other faculty. Cann and his colleagues developed *Greening Across the Chemistry Curriculum*[13] to provide "modules in green chemistry to insert into existing courses across the college chemistry curriculum."[14] The modules expose students to real-world state-of-the-art examples of green chemistry as part of the mainstream college curriculum. There is also an interest to put green chemistry into the business side of courses. Web-based modules have also been developed for the following existing chemistry courses: general, organic, inorganic, physical, environmental, industrial, and polymer chemistry, as well as toxicology and biochemistry. Each of the Web-based modules has three parts:

1. "The module": A green chemistry topic is discussed in class, and then the instructor directs the students to visit the Web page to read and study the material.
2. "Notes to Instructors": Suggestions are provided to aid instructors in determining where a module could be used in a particular course and other courses.
3. "PowerPoint Presentation": Instructor can use PowerPoint presentations to present the material, and students can use them as notes.

The project had funding from the Camille and Henry Dreyfus Foundation Special Grant Program in the Chemical Sciences, the ACS/EPA Green Chemistry Educational Materials Development Project, and the University of Scranton.

Lastly, Cann featured Colin Baird's *Environmental Chemistry*[15] as an example of a text that has green chemistry integrated throughout every chapter. In addition, the preface is an introduction to green chemistry, atom economy, and the synthesis of ibuprofen.

The next panelist, Dr. Eric Beckman from the University of Pittsburgh, focused on chemical engineering and sustainability in his presentation. Beckman began by discussing the chemical engineering community's reluctance to incorporate green chemistry into their curriculum. He stated that most chemical engineers think that "basic fundamental chemical engineering is green engineering, end of story, on to the next thing." To survey the reality of chemical engineering for himself, Beckman analyzed each principle of green engineering. He found some items were consistent with current processes and some were not. According to Beckman, a major deficiency with "both chemistry and chemical engineering curricula is that we don't worry about product design very much." He added that emphasizing product design is important in overall design paradigm. Issues with the deficit in product design include:

- "The majority of students trained in chemistry and chemical engineering who enter industry will work in product-related functions, yet few receive formal training in product design and development."
- "It is not clear to many of our students that they will one day have customers, that this is a good thing, and that perhaps one should interact with the customers."
- "In academia, all of our 'products' are single component and 99 percent plus pure."
- "If we are not currently teaching product design, how then do we add sustainability as a constraint?"

Beckman emphasized that the convergence of chemistry and engineering is needed to accomplish real green design. Beckman cited an article from the *Journal of Business Research*[16] on how to achieve sustainable product design. The article features three approaches:

1. Eco-redesign (E-) = short term, modify current design, reduce waste, preserve business as usual—the "low hanging fruit";
2. Eco-innovation (E+) = longer term, reinvent ways and means to provide benefits to customers; and
3. Sustainable technology innovations (E++) = emerging or unproven technology to provide through inherently different mechanisms; radical technology change.

After deciding on an approach to teaching product design, metrics must be used to gauge progress.

In closing, Beckman explained what he thinks is needed to teach a chemical product design course: The course should be team taught and available to multiple disciplines, should use sustainability as a constraint, should use validation tools, and should consider the voice of the customer as well as adequate product performance and price.

The next panelist was Dr. Kathryn Parent, from the Green Chemistry Institute (GCI), who discussed American Chemical Society (ACS) resources available for green chemistry education. Parent explained that GCI's mission is "advancing the implementation of green chemistry and engineering principles into all aspects of the chemical enterprise," including education. In answering that charge, GCI and ACS have developed an aggregate of tools, materials,

[13] http://academic.scranton.edu/faculty/CANNM1/dreyfusmodules.html.

[14] http://academic.scranton.edu/faculty/CANNM1/dreyfusmodules.html.

[15] Baird, C., and M. Cann. 2004. *Environmental Chemistry*, 3rd ed. New York, NY:WH Freeman. Available at http://bcs.whfreeman.com/envchem3e/.

[16] Fuller, D., and J. Ottman. 2004. Moderating unintended pollution: The role of sustainable product design. *Journal of Business Research*. 57(11): 1231-1238.

and programs geared toward greening chemistry education. GCI is attempting to make these changes through the development of new courses or the incorporation of content into existing courses, research, and extracurricular activities, such as student affiliates, conferences, workshops, symposiums, and ACS summer schools.

According to Parent, "In 2001 there were no educational materials on green chemistry available to educators beyond technical reference books. By 2005 GCI in partnership with ACS Education had produced six green chemistry publications for chemical educators. Over 1,000 copies per year are distributed to customers. GCI receives requests for green chemistry educational materials from faculty around the world." Parent displayed a list of available education materials:

- *Chemistry in the Community*—A high school textbook;
- *Introduction to Green Chemistry*—A high school unit text;
- *Chemistry in Context*—An undergraduate textbook;
- *Real-World Cases in Green Chemistry*—An undergraduate seminar text;
- *Green Chemistry: Innovations for a Cleaner World*—A companion video to Real-World Cases in Green Chemistry;
- *Greener Approaches to Undergraduate Laboratory Experiments*—An undergraduate laboratory experiment manual;
- *Green Chemistry: Meeting Global Challenges*—A DVD of conference presentations;
- *Going Green: Integrating Green Chemistry into the Curriculum*—A how-to resource for faculty; and
- Online resources
    - ACS Green Chemistry Institute, http://www.greenchemistryinstitute.org
    - ACS Education Division, http://www.chemistry.org/education/greenchem
    - Annotated bibliography, http://chemistry.org/greenchem/bibliography.html

In addition to the resources listed above, ACS continues to develop new resources such as new textbooks infused with green chemistry; business school case studies being conducted to emphasize the connection between green chemistry and economics; and other user driven tools. Parent concluded that students are "our greatest resource in green chemistry education and developing them should be our key goal."

Dr. Richard Wool from the University of Delaware gave the last presentation of the panel and of the first day of the workshop. Wool discussed his senior undergraduate course "Green Engineering Out of This World." The class typically consists of about 30 students that are split into eight to ten Web teams. The students learn the basic tools of green engineering systems and how to do the adequate analyses using sustainability issues as the subjects. The class structure is:

- Web teams—Major environmental drivers;
- Global sustainability issues—National Academy of Sciences;
- Course work—*Green Engineering: Environmentally Conscious Design of Chemical Processes* by David Shonnard and David Allen;
- EPA Web sites;
- Individual projects;
- Peer review; and
- Unintended consequences—Green court.

Required reading for this course includes the *Green Engineering Tutorial: Environmentally Conscious Design of Chemical Processes* by Allen and Shonnard and *Bio-Based Polymers and Composites* by Wool.

BREAKOUT SESSIONS

On the second day of the workshop, breakout sessions allowed participants to delve deeper into the issues surrounding green chemistry and engineering. Workshop participants were assigned to breakout groups, and the results of those breakout sessions that corresponded with the tools and materials for green chemistry and engineering education are summarized below.

What Materials, Programs, and Tools Are Needed?

The participants in this breakout group believed that any tools, materials, or programs for green chemistry and engineering would be most beneficial if they were targeted at the undergraduate level and possibly the industrial level. The participants identified incorporating green chemistry into mainstream textbooks as one way to overcome barriers associated with teaching green chemistry and engineering to chemistry and chemical engineering majors, as well as other science, engineering, and nonscience majors. The participants thought this technique was a reasonable way to engage students and raise awareness about green chemistry and engineering. In addition, a global motivation document could be used to attract new audiences by presenting an overarching view of the main issues in green chemistry and engineering. This technique will be beneficial only if the book is not ignored.

Other tools, materials, or programs needed to complement current green chemistry and engineering educational resources include:

- Introduction or capstone to design course for engineers and scientists;
- Integrated laboratory and lecture courses;
- Seminar courses on modern topics in green chemistry;
- Comprehensive centralized Web-based resources; and
- Assessment tools for undergraduates.

Examples of recent efforts that provide a sufficient starting point for green chemistry are the ACS efforts that Parent presented and the University of Oregon's Greener Education Materials for Chemists (GEMs) that Haack presented. Management, coordination, and funding of efforts are required for future adequate expansion. One note of caution is that not all tools can be adapted for the targeted educational purposes. An example of this is Building for Environmental and Economic Sustainability (BEES), software developed by the National Institute of Standards and Technology (NIST) directed to aid in selecting cost-effective, environmentally friendly building products using green principles. Unfortunately, this tool is applicable only to construction.

Barriers to using current green chemistry and engineering materials were also identified. Chemical engineering has defined a set of core principles of which green engineering is considered to be outside the scope of these core concepts. Professors are expected to achieve higher learning curves for students and have to factor in the time constraints of adding lessons to an already full course curriculum. Untested case studies and examples of green chemistry and engineering could have unintended consequences. Any unintended consequences related to green chemistry and engineering could dampen credibility, foster distrust of green chemistry and engineering, or discourage participation and support of green chemistry and engineering among professionals and students.

In general, the majority of the group participants agreed that infusing green chemistry and engineering into textbooks and improvement of textbooks by professional societies are ways of enhancing curricula. They also agreed that although it may not seem difficult to integrate green chemistry and engineering into textbooks, the efforts will not be successful without the support of textbook authors and a seal of approval from professional societies.

What Is Needed to Achieve Interdisciplinary Approaches?

In this breakout session the group addressed issues regarding interdisciplinary educational approaches. Some barriers to interdisciplinary collaboration are:

- Internal issues within institutions or organizations;
- External support mechanisms; and
- Recognition of expertise.

In interdisciplinary endeavors, department chairs have a number of administrative barriers that cause them to be reluctant to engage in partnerships: (1) how to distribute expenses for necessary materials across departments; (2) how to allocate the time commitment of faculty across departments; (3) intellectual property issues; and (4) the burden of adding electives in addition to core coursework. To overcome barriers in interdisciplinary endeavors, department must see the value in collaboration. A reward system to motivate these partnerships may encourage interdisciplinary collaboration and encourage departments to see the value in collaboration outside their departments, but other value propositions must also be identified.

The presence of cultural barriers that impact interdisciplinary approaches was also discussed. The language of chemistry, corporate influence on chemistry and chemical engineering, and differences in processes and approaches in chemistry versus chemical engineering are three of the cultural barriers. Chemistry has a very unique language that other disciplines do not always easily comprehend. A concentrated effort to speak one another's languages could diminish the language barrier. The different approaches and processes in chemistry versus chemical engineering is apparent since chemistry focuses on pure science and chemical engineering focuses on applied science. The focus on innovation in chemical engineering may allow for an easier integration of green principles.

Interdisciplinary approaches tend to be viewed differently by industry and academia. A high interest in interdisciplinary collaboration has been shown in industry. The participants believe that students may like working in teams for research purposes but dislike working in teams on graded classroom projects.

At the end of this breakout the participants agreed on the following as possible actions to address the interdisciplinary issues:

- Develop a framework for funding;
- Increase awareness and information sharing between disciplines;
- Develop a reward system to recognize good practices; and
- Develop leadership from key faculty across disciplines.

4

Where Do We Go from Here?

During the first day of the workshop, the discussion centered on current green chemistry education accomplishments. During the second and final day of the workshop, the participants brainstormed about the future direction of green chemistry education. Workshop organizer Paul Anastas stated that this session aimed to capture the best thoughts, strategies, and tactics about green chemistry education capabilities and motivations. The session began with talks by four main speakers, followed by five panel speakers.

MAIN SPEAKERS

The first main speaker in this session was Dr. F. Fleming Crim from the University of Wisconsin. Dr. Crim has multiple perspectives since he is a college professor, chair of the Chemical Sciences Roundtable, which was the birthplace of this workshop, and a member of the American Chemical Society's Committee on Professional Training (CPT). Crim first presented his CPT perspective and the role CPT can play to facilitate green chemistry into curricula. The general goals of the CPT are:

- The promotion of excellence in chemistry education and in professional training of chemists;
- The gathering and dissemination of information that maintains and improves the quality of chemistry education beyond the secondary level;
- The facilitation of refinements and changes in chemistry education that reflect the modern and evolving face of the discipline; and
- The maintenance and enhancement of an effective approval procedure for undergraduate chemistry programs that benefits the programs, students, and employers by providing the greatest return on their efforts and those of the committee and staff.

Since CPT plays a role in the ACS approval program for chemistry undergraduate programs, CPT would like to facilitate bottom-up change to implement the use of green chemistry in undergraduate curricula. CPT's role in bottom-up change would first be to define an excellent green chemistry education program and then let the community respond.

Crim mentioned that there are a few details to note about a curriculum development process. First, there are many good competitive ideas in the marketplace and green chemistry is competing with nanoscience, chemical biology, and others. Therefore, more publicity and advocacy for green chemistry may be needed to bring it to the forefront of other ideas. Crim emphasized faculty acceptance as another issue. He said that most faculty members seem to be receptive to green chemistry in the curriculum but are overwhelmed with an already full curriculum. Because CPT is working on integrating more flexibility into the curriculum, Crim encourages advocates of green chemistry to become involved in the CPT process.

According to Crim, the three most important things to make changes occur are materials, materials, and materials. Crim suggested that if infusing green chemistry into the curriculum broadly is the preferred approach, low-entry barrier and bite-sized increments are needed to appeal to overwhelmed faculty. On behalf of CPT, Crim offered the CPT newsletter as a forum to provide green chemistry examples or links to green chemistry materials to the community.

Crim spoke next from the R1 perspective. According to Crim, there are a few things that the community could do to change the opinions of R1 institutions. Again, providing materials is essential since R1 faculty do not feel that they have time to create new materials. Second, highlighting intellectual opportunities within green chemistry and advocating funding for green chemistry research is needed because both funding and intellectual opportunities drive R1 research. Third, Crim thought that "teaching the organizing principles of chemistry, science, and physical science within the context of green chemistry is extremely important and also appeals to people who are not necessarily in the mainstream of green chemistry." He said that when advocating for green chemistry, the argument should be presented in a fashion that suggests that green chemistry rests on the core organizing principles of chemistry. Fourth, Crim pointed out that some factions of the community perceive green chemistry as soft science. Crim suggested this may come from advocating that green chemistry and engineering is socially responsible. This perspective may not appeal to those at the "core" of chemistry. Talking about the organizing principles and the intellectual opportunity of green chemistry may also be more appealing to those who reject the social argument.

The next speaker in this session was Dr. Cliff Davidson from Carnegie Mellon University's Environmental Institute. Davidson divided his talk into four topic areas: (1) skills and attitudes that future engineers will need; (2) environment across the curriculum initiative; (3) case studies in green engineering; and (4) center for sustainable engineering: A three-university consortium comprising Carnegie Mellon, University of Texas at Austin, and Arizona State University.

Davidson believes that to move green engineering forward, engineers need to "go beyond reductionist thinking, where each part of a complex system is considered separately—emphasize the emergent properties of the whole." The skills and attitudes future engineers will need are:

- Sensitivity to the environment—Exposing engineering students to issues of the environment is becoming of increasing importance. Environmental engineers ponder whether education can transform students who are not environmentally sensitive or whether it is the responsibility of the environmental engineering community to proactively recruit students that are environmentally sensitive into engineering.
- Sensitivity to human needs—Educating students about sustainable engineering to increase humanist interests is not the norm but may need to be taken into consideration in engineering education.
- An ethical foundation—According to Davidson, the engineering practice lacks a strong environmental ethic as a basis for decisions, yet sustainability issues are becoming a part of engineers' ethical responsibilities. Many engineers are faced with scenarios in which they have clients who are not supportive of environmental preservation. Davidson believes that exposing students to this dilemma in class will better prepare them for future challenges.
- Understanding of natural systems—An understanding of natural systems (i.e., ecosystems) from the life sciences, physics, and chemistry, perspectives is necessary for engineers of sustainable design.
- Understanding of social systems—Engineering decisions are made in the context of societal systems (e.g., legal, economic, and political), but most engineers do not have expertise in these areas. Possessing knowledge of these social systems could allow engineers to become more politically and socially sensitive, as well as help scientists promote important agendas, such as sustainability.

Davidson next spoke of the Carnegie Mellon initiative "Environment Across the Curriculum." The initiative goal is to introduce environmental modules into nonenvironmental courses. Davidson provided many examples of successful environmental modules being integrated into departments across other college campuses

Lastly, Davidson discussed the Center for Sustainable Engineering, a team effort from Carnegie Mellon University, Arizona State University, and University of Texas at Austin with support from the National Science Foundation and EPA. The center's goal is to "develop and implement activities to enhance education in sustainable engineering at colleges and universities around the world."[1] Workshops for engineering faculty who would like to add sustainable engineering to courses began in July 2006.

A second activity is a Web site called Bookbuild that is a partnership between the three universities and Pearson/ Prentice Hall. Bookbuild will be a global hub for faculty to submit and share lecture materials, notes, slides, handouts, and other engineering educational materials. All materials submitted to this Web site will be subject to peer review.

Another activity for the center is a benchmark assessment of the status, including materials, of sustainable engineering activities in U.S. engineering departments. Davidson left the audience with the message that changing engineering courses is necessary to teach engineers green skills and attitudes.

Next, Dr. Terrance Collins, director of the Institute for Green Oxidation Chemistry at Carnegie Mellon University, provided his perspective on "Where do we go from here?" He explained that the goal of the institute is to perform world-leading research in green oxidation analysis. The institute is very important in green chemistry because the first green chemistry course was taught there in 1992, and it continues to be offered to upper level undergraduates and beginning graduate students. It is clear that green chemistry has been on Collins's and his colleagues' minds for quite some time. Collins has trained students who have won many awards, including the Alexander von Humboldt Postdoctoral, Beckman, Goldwater, and Hancock, and in his opinion, these people are the next leaders in green chemistry.

According to Collins, "sustainability is the single most important challenge for our civilization for at least the next 100 years." Collins stated that the cause of our sustainability problem is science and technology, which is controversial for the universities and disciplines since they are also re-

[1] http://www.csengin.org/.

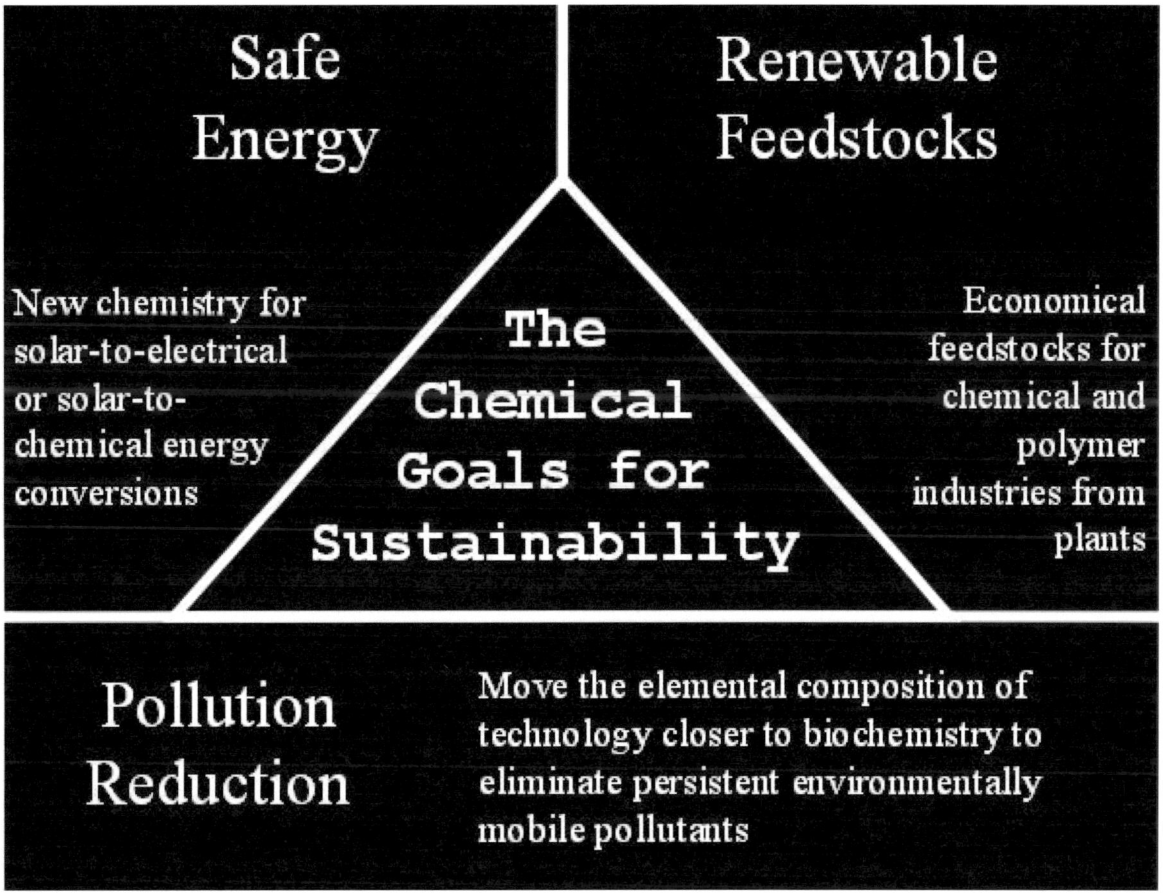

FIGURE 4.1 Chemical goals for sustainability. SOURCE: Collins, T. 2005. Where Do We Go from Here to Green Our Civilization through Science? Presentation at the National Academies Chemical Sciences Roundtable Green Chemistry and Engineering Education Workshop. November 8, 2005.

sponsible for developing the science and technology. Although various federal funding agencies and private foundations support the green chemistry efforts at Carnegie Mellon, many universities are not willing to risk losing funding in order to address sustainability issues. Collins believes that any university that wants to be an honest actor in sustainability must be prepared to deal with controversy. Collins pointed out, however, that leadership has come from of people such as Paul Anastas and the workshop audience to drive universities toward the important issue of sustainability.

Collins paraphrased Hans Jonas's *The Imperative of Responsibility: Finding an Ethics for the Technological Age*[2] by saying that "all previous ethics have been based on the premises that the human condition is determined by the nature of man and the nature of things." He gave the analogy of living in ancient times where the people thought that what they did impacted only the people they came in contact with. During those times, people did not think that what they were presently doing could impact people in the future. That premise has changed. What we are doing today is going to impact the welfare of many future generations because of our power over the ecosphere through science and technology. Collins believes that recognizing this future impact principle and building upon it is essential and, therefore, green chemistry is essential.

Next Collins presented the chemical goals for sustainability (see Figure 4.1).

The first goal discussed was safe energy. In Collins's opinion we do not have an energy problem; we have an energy policy problem. Collins believes that if a sustainable technology base is developed, the energy problem would be nonexistent, and we will have safe energy. According to Collins, safe energy equals solar energy, and we need new chemistry for solar-to-electrical or solar-to-chemical energy conversions to achieve this goal.

The second chemical goal for sustainability that Collins presented was renewable feedstocks. Economical feedstocks for chemical and polymer industries from plants are needed.

[2]Jonas, Hans. 1984. *The Imperative of Responsibility: In Search of an Ethics for the Technological Age*. Chicago, IL: University of Chicago Press.

Collins believes that we can get the things we need out of recently dead plant matter rather than fossilized plant matter, and this is an active area of industrial research.

The final chemical goal for sustainability was pollution reduction. Collins believes that this can be done by moving the elemental composition of technology closer to biochemistry to eliminate persistent environmentally mobile pollutants. Collins's research group is focused on this goal. In particular, Collins promotes solar Stirling engines. Robert Stirling invented the Stirling engine in 1816. The engine works on a heat differential and because it does not have explosions in the pistons or make noise, has been used in nuclear submarines.

Lastly, Collins discussed the enormous stakes of failing to address toxicity and ecotoxicity. He first highlighted laboratory research that exposed pregnant rats to a mixture of DDT and vinclozolin. The male offspring experienced severe reproductive damage up to four generations later. He also noted the research revealing developmental impairment due to lead toxicity. Lead toxicity is still a problem today in places where lead is persistent in drinking water. Collins stressed that failing to address environmental toxicity issues could have severe repercussions in the future.

The final main speaker in this session, Dr. Julie Zimmerman from the EPA and the University of Virginia, followed Collins's concerns about toxicity and focused on how design decisions impact cost, waste, and the environment. Zimmerman stressed the importance of big picture questions, such as investments, time, energy, resources, money, and potential realized benefits, rather than just design questions. To impact all of these elements, Zimmerman highlighted three steps to change design procedures: (1) optimize the existing solution; (2) reengineer the system; and (3) redefine the problem.

According to Zimmerman, the same challenges occur when designing a new curriculum and designing new products and processes. Zimmerman stressed that introducing business, social science, service, production, and design at an early stage will help move toward a more integrated curriculum with multidisciplinary teams. Zimmerman mentioned 10 disciplines for incorporating sustainability in product design:

1. Research and extract engineering,
2. Materials science,
3. Mechanical engineering,
4. Chemistry,
5. Chemical engineering,
6. Manufacturing engineering,
7. Civil engineering,
8. Environmental science,
9. Social science, and
10. Policy making.

Zimmerman believes that the multidisciplinary approach also embraces a necessary holistic approach. One way she is trying to achieve this is through the EPA P3 (People, Prosperity, and the Planet) competition that asks students to identify what they see as a challenge to sustainability and propose a scientific, technical, or policy solution The proposals are peer reviewed, and outstanding applicants are given $10,000 grants to perform their proposed work. Grantees are required to develop an interdisciplinary team and to quantify the benefits of their design environmentally, economically, and socially. At the end of the $10,000 academic award, students participate in a second round of competition in Washington, DC. Six winners from the second round are awarded phase two funding. The funding gives the winning grantees $75,000 grants to further develop their designs and move toward commercialization. The program has spawned entrepreneurship and innovation, such as student teams developing courses and start-up companies.

For those who work on college campuses, Zimmerman explained how there are many opportunities to integrate sustainability into the physical infrastructure of the campus. The opportunities include transportation decisions, where to put new buildings, energy, and managing hazardous waste, and these particular influences can be measured, which gives a means for measuring the impact of sustainability decisions.

Zimmerman devoted the final portion of her talk to the intellectual pipeline of people. Zimmerman feels that this is the time to embark on the issues of ethnic and gender diversity in the workforce. If this issue is addressed now, when sustainability has the attention of the scientific community, we can gain the benefit of a diverse workforce that is engaged in and trained in sustainability.

PANEL SPEAKERS

Dr. Linda Vanasupa from California Polytechnic State University was the first panelist of this session. Vanasupa's discussion focused on curricula stemming from scientific discovery and the human dimension of designing curricula. When designing curricula, the ultimate goal is to produce scientists, engineers, technologists, and practitioners who are capable of practicing or applying sustainable solutions. These solutions should also reflect the society they serve.

Although there are adequate numbers of students in the pipeline, ideas of how to attract the people that reflect society (i.e., societal demographics) and of how to retain them in science and engineering programs are necessary. For example, there are a number of studies about why women drop out of engineering. The basic reason for women leaving engineering is because they do not see engineering as relevant to their life goals. Caltech received a grant from the National Science Foundation that they are using to attract a diverse population of applicants. One tactic that was used was an e-mail message that was sent to high school students in California, but the message did not aid Caltech in achieving its goal of attracting a diverse population of applicants. It attracted only two female students. The message was redesigned to appeal to a different audience. Vanasupa highlighted several best practices to help retain all students:

- Systems thinking;
- Meaningful context;
- Integration of support subject domains;
- Interaction with faculty as coaches;
- Active learning and design;
- Connection with peers;
- Reflection and self-assessment of learning; and
- Emphasis on the American Board for Engineering and Technology's (ABET) "other" design constraints.

The next panel speaker, Dr. John Leazer of Merck Co., discussed green chemistry from the pharmaceutical industry perspective, where innovation drives the use and implementation of green chemistry. Leazer sees green chemistry as a contributor to industry goals of innovation, efficient processes, and integrated business flow through several efforts. He explained how the efforts comprise demanding exceptional chemistry, teamwork, integration of discovery and manufacturing objectives, cost-effective processes, enhanced safety, and quality performance. Leazer emphasized that green chemistry is a business advantage because it can be key to achieving other initiatives, such as:

- Product optimization;
- Energy conservation;
- Lean manufacturing;
- Operational excellence;
- Sustainability;
- Technical leadership;
- Enhanced productivity; and
- Get it right the first time.

Merck is one company with leading activity in green chemistry. The company received the Presidential Green Chemistry and ICHEME Astra Zeneca awards for focusing on the 12 principles of green chemistry and for solid efforts to implement green chemistry.[3] Buy-in at the highest levels of Merck influences the status of green chemistry at Merck. Sponsorship and buy-in at the highest levels of the research and manufacturing divisions have made implementation of green chemistry successful at Merck. To substantiate these goals Leazer quoted Paul Anastas and John Warner as saying, "The use of auxiliary substances should be made unnecessary wherever possible and innocuous when used." Examples of Merck's efforts to use green chemistry are:

- Developing a collaborative green chemistry effort between R&D and manufacturing;
- Creating of a green chemistry advocate process research position;
- Joining the ACS Green Chemistry Pharmaceutical Roundtable in 2005;
- Fostering early stage environmental process review to identify opportunities for waste minimization, recycling, and process streamlining prior to production;
- Developing a process research mission statement that embodies principles of green chemistry:
    - "Definition and demonstration of practical, scalable, efficient, cost effective and environmentally benign chemical processes."
- Starting initiatives in both chemical and biocatalysis research:
    - Reaction optimization;
    - Atom economy, minimal waste, minimal metal usage, fewer protecting groups, no chiral auxiliaries; and
    - High throughput screening and miniaturization that reduce reagent and solvent usage.
- Using supercritical fluids rather than traditional chromatography to reduce solvent and energy use; and
- Providing education and training through seminars and symposiums.

Leazer explained how these efforts motivate chemists to be cognizant of waste and total mass balance of processes, opportunities for recycling, less hazardous reagents, and solvent minimization.

In closing, Leazer emphasized several points for green chemistry education. First, Leazer emphasized the importance of having a thorough understanding of chemistry with green chemistry being taught in addition to core competences. Second, he underscored the need to more closely align academia and industry. Third, he stated the need for public outreach initiatives because "they (the public) don't hear chemical itself, they hear toxic chemical." Lastly, Leazer emphasized the empowerment of critical development technologies to offset green chemistry investments.

Dr. James Mihelcic from Michigan Technological University was the next panelist to speak. Focusing on the theme of partnerships, he divided his talk into three sections: (1) the integration of green chemistry and engineering with sustainability; (2) the global perspective, and (3) the issues of diversity. Mihelcic stated, "We need to teach our students how green chemistry and engineering will generate wealth for society and industries they work for, and importantly, how it will make our nation globally competitive." Mihelcic continued that the nation's global competitiveness will ultimately depend on how schools, colleges, universities, and other education providers from (precollege through postdoctoral training) develop and refine human resources.

Mihelcic spoke in detail about Michigan Tech's National Science Foundation IGERT doctoral training grant that follows

[3]For more information on the 12 principles of green chemistry and engineering, see the following Web site: http://www.chemistry.org/portal/a/c/s/1/acsdisplay.html?DOC=greenchemistryinstitute\whatare.html.

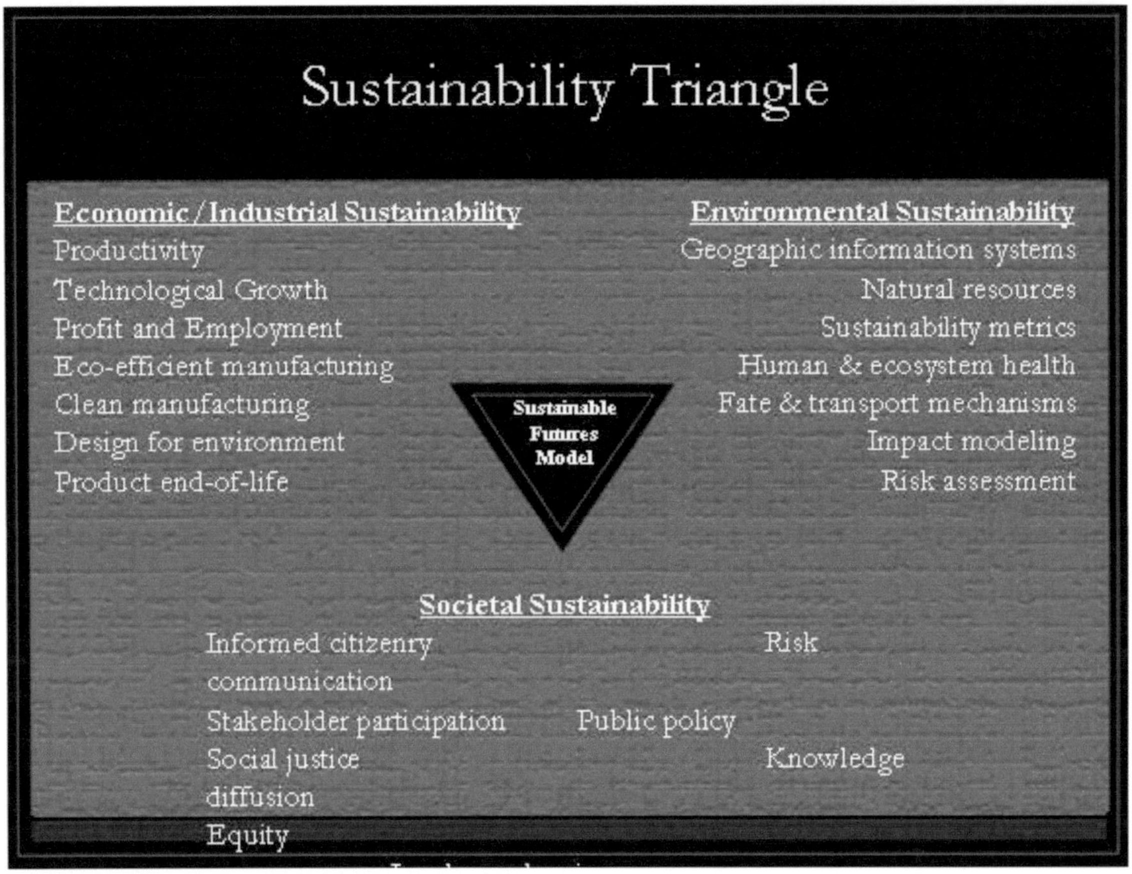

FIGURE 4.2 The sustainability triangle. SOURCE: Mihelcic, J. 2005. NAS Green Chemistry/Engineering Workshop: Where Do We Go from Here in Green Chemistry and Engineering Education? Presentation at the National Academies Chemical Sciences Roundtable Green Chemistry and Engineering Education Workshop. November 8, 2005.

a sustainable future model. The model has economic and industrial, environmental, and societal components (see Figure 4.2). In addition to teaching students about the three components, Mihelcic encouraged teaching students the business aspect of education as well.

Next Mihelcic focused on the opportunity to educate, as well as retain, students with global thinking because the developing world's problems are related to water, soil, agriculture, forestry, and fisheries rather than the manufacturing in the modern world. Mihelcic displayed a quote from National Academy of Engineering President William Wulf: "We need to understand why in a society so dependent on technology, a society that benefits so richly from the results of engineering, a society that rewards engineers so well, engineering isn't perceived as a desirable profession. . . . Our profession is diminished and impoverished by a lack of diversity." Mihelcic cited Michigan Tech's and Southern University and A&M College's joint engineering and public policy Ph.D. programs, supported by National Science Foundation IGERT and REU grants, as an example of global education. The partnership relates sustainability issues between the Great Lakes in Michigan and the Mississippi River in the South. The collaboration allows diversification of participating faculty members at each school by offering joint appointments at both universities, program offerings, and students.

The next panel speaker was Dr. Jorge Vanegas, formerly of Georgia Institute of Technology. Vanegas has a background in architecture, construction, and civil engineering that provided a very different perspective from chemistry and engineering. He spoke about strategies and approaches necessary to make green efforts happen based on Georgia Tech's green efforts. The strategies and approaches he suggested were a mission statement, a comprehensive philosophical approach, a goal, a long-term plan, support from the top, results, campus-wide integration and coordination, appropriate infrastructure, and money:

- A mission—Through many efforts, Georgia Tech's mission now includes sustainability.
- A comprehensive philosophical approach—This approach includes learning in the classroom, discovery in the research laboratory, and active management of the campus.

- A goal—Students, faculty, and staff need to understand their respective roles in creating a more prosperous and sustainable society to be in line with the vision.
- A long-term plan—The plan must begin with engaging the faculty by creating a grassroots-driven vision. Curricular innovation for all students in every major must be achieved. Implementing reliable campus practices can build trust and credibility. Being in possession of a passionate advocate could help move your vision forward.
- Support from the top—Institutional commitment and an institute-wide agenda will encourage the support needed from top-level institutional officials.
- Results—Two Georgia Tech faculty members who have collaborated for more than 15 years on sustainable chemical processes are among the winners of 2004 Presidential Green Chemistry Challenge Awards. Other examples of results could be published journal articles or recognition from the scientific community.
- A campus-wide mechanism for integration and coordination. The Institute for Sustainable Technology and Development at Georgia Tech serves in this capacity. The institute encourages activities in education and research and managing the campus and stimulates activities with other universities and industry.
- Appropriate infrastructure. For example, at Georgia Tech, interdisciplinary research neighborhoods have been created to enhance collaboration and innovation.
- Effective strategies and tools are necessary in education, research, and campus wide to influence change.
    - o Education
        - ■ Strategies for education include using existing curricula and integrating new concepts into major programs and general studies.
        - ■ Tools needed for education include special initiatives, curriculum committees, academic support, accreditation self-studies, and assessment.
    - o Research
        - ■ Strategies for research include enhancing existing R&D programs and fostering new R&D programs for faculty development.
        - ■ Tools for research include research facilities, faculty recruitment, endowed chairs, "seed" funds, centers, and initiatives.
    - o The campus
        - ■ The campus strategy is to weave concepts of sustainability into policies and procedures.
        - ■ Tools for the campus include the campus master plan, operations, and purchasing guidelines.
- Money—In addition to funding, Georgia Tech has 21 endowed chairs and professorships that are related to sustainability, which makes sustainability an integral part of its capital campaign.

Lastly, Vanegas stated that it was necessary to "walk the walk." According to Vanegas, Georgia Tech has been doing green chemistry for several years. On the construction, engineering, and architecture side of things, Georgia Tech is demolishing parking lots to create green space, providing alternative campus transportation, constructing new buildings that are LEED[4] certified, and planting trees as part of a campus-wide tree canopy renewal. Another part of walking the walk is in investments. For example, Georgia Tech is investing in focused research programs such as organic photovoltaics, the Center for Bio-inspired Design, and closed-loop production systems. Vanegas encouraged the audience to *carpe diem*, *carpe noctum*, and *carpe momento,* which means seize the day, seize the night, and seize the moment.

Dr. Liz Gron from Hendrix College was the final workshop panelist. Gron talked about educating green citizens. She began her presentation by showing who the green community is now and who it could be. Gron said that it is possible that implementation of a green curriculum could expand the current community, many of whom were at the workshop, to include all undergraduate majors. Of the 2.5 million first-time college freshmen, 70 percent had no interest in scientific or professional studies, 12 percent were interested in professional studies, 7 percent were interested in biology, and 11 percent were interested in the physical sciences. Although focus needs to be put on retaining the students interested in the physical sciences (only 0.6 percent of the 11 percent interested in physical science graduated in chemistry or engineering), Gron believes that the vision needs to be expanded to include green scientists and professionals.

Gron then discussed the Green-Soil and Water Analysis at Toad-Suck (Green-SWAT) Laboratory Program at Hendrix College. This program teaches green, analytical, and environmental chemistry to introductory students, which Gron feels is working because it dispels the exclusivity of environmental chemistry. This is achieved by teaching introductory students, cultivating environmentally and scientifically "savvy" students, and instilling a green ethic in students. The results of these efforts, in turn, influence the students' future professional, business, or personal choices.

Gron identified two challenges for teaching green chemistry but also provided ways to overcome these challenges. Gron recommends the following actions to move green ideas from a local audience to a global audience and to overcome the challenges:

- Create and encourage local educational initiatives
    - o Small and large efforts, including outreach activities and/or whole majors
- National initiatives
    - o Start-up funding to encourage the smaller companies to invest in green chemistry principles

---

[4]The LEED (Leadership in Energy and Environmental Design) Green Building Rating System® is a voluntary, consensus-based national standard for developing high-performance, sustainable buildings sponsored by the U.S. Green Building Council (www.usgbc.org).

- Disseminate information
  - o Forums and symposiums;
  - o Journal and newspaper articles; and
  - o Textbooks.

Gron concluded that the current community must be prepared to encourage people into green chemistry and everything else will follow.

BREAKOUT SESSIONS

On the second day of the workshop, breakout sessions allowed participants to delve deeper into the issues surrounding green chemistry and engineering. Workshop participants were assigned to breakout groups, and the results of those breakout sessions that corresponded with the session on "Where do we go from here?" in green chemistry and engineering education are listed below.

Creating Incentives, Removing Impediments

In this breakout session, participants explored green chemistry and engineering incentives, impediments, and ways to remove the impediments in both academia and industry. The absence of a clear vision statement and the lack of scientists in the policy-making arena pose significant barriers for both academia and industry. The participants acknowledged that potential regulatory barriers in industry exist. There was also a general feeling that industry will not adopt green principles unless there is market demand. In academia there are inadequate numbers of faculty trained in green chemistry and engineering, a lack of available tools, a competition between green and traditional coursework, a lack of time for approval or implementation, and tenure criteria not viewing green chemistry as a rigorous discipline.

The group was able to identify incentives for academia and industry. For industry regulations, ISO-like certification and a viable market could act as incentives for companies to adopt green processes. Other ideas for incentives for academia and industry that could potentially raise awareness and decrease skepticism included:

- Presenting awards for excellence in green chemistry and engineering education, possibly connected to Green Chemistry Challenge awards;
- Having more leaders in green chemistry and engineering speak at general conferences and meetings;
- Developing materials that explain the relevance of green chemistry and engineering to other areas, such as policy, economics, and public health;
- Providing business cases based on real examples to encourage industry;
- Highlighting green principles in university and industry wide publications;
- Connecting green chemistry and engineering to major sustainability issues;
- Indicating the need for green chemistry or engineering experience in employment announcements;
- Utilizing ACS for proposing short courses in green chemistry and engineering; and
- Recruiting ACS members to buy into green chemistry and engineering by teaching it and speaking about it.

Green Chemistry and Green Engineering in Future Curriculum

The participants identified ways that green chemistry and engineering could be incorporated into future curricula. Most participants believed that the following items are needed to implement green principles into curricula:

- Provision of high-quality materials and resources, such as:
  1. Improvements to current materials and resources by replacing lessons in books that incorporate green chemistry and engineering;
  2. An overall intellectual framework for green chemistry and engineering modules;
  3. Seminars centered on green chemistry and engineering; and
  4. Published articles highlighting green chemistry and engineering in major academic journals.
- Development of interdisciplinary interactions by finding simple access points in other disciplines where green chemistry and engineering are applicable;
- Recognition through awards; and
- Changes to current curricula to accommodate green chemistry and engineering, such as:
  1. Offering green chemistry and engineering electives; and
  2. Having laboratory managers incorporate green chemistry and engineering concepts into laboratory experiments at all levels.

The subject of developing specific degree tracks in green chemistry and engineering raised a number of differing views in the breakout session. Some participants believed that green chemistry and engineering need to be an integral part of all good degree programs and taught in an interdisciplinary manner at the graduate level. However, these participants thought that a specific degree track would limit a degree candidate's career opportunities. There were other participants that supported the idea of a specific degree track. The group suggested that a master's-level program leading to a Ph.D. degree could fill a niche that a Ph.D. program alone cannot fill. Most participants agreed that an undergraduate degree was not appropriate because Bachelor of Science graduates are trained to be generalists; graduate degree programs are more specialized.

5

Overarching Curricula and Implementation Ideas

In addition to the presentation of curricula being used and developed at all educational levels, the workshop allowed the attendees to react and discuss ideas about green chemistry and engineering education. Five themes that came up throughout the two days and in multiple education categories were (1) marketing; (2) green curricula; (3) research extensive universities (R1s); (4) business education; and (5) green ethics. This section highlights these overarching ideas.

PROMOTING GREEN CHEMISTRY AND GREEN ENGINEERING

The marketing of green chemistry and green engineering education efforts within an organization and to the public are critical for the growth of the fields. Linda Vanasupa, California Polytechnic State University, San Luis Obispo, is not so sure that if green chemistry is incorporated into the curricula students will automatically be drawn to the idea; the human dimension must be considered. Green educators are excited about new curricula but also need to balance that with the issue of what students want. The educators have to overcome the perception that green chemistry and engineering are niche topics rather than state-of-the-art science and therefore can attract and retain the best and brightest students.

Several of the attendees expressed ideas of how to improve marketing of green education. Steve Howdle from the University of Nottingham explained that the reason for the low enrollment rate for the undergraduate green program may be as simple as where the green chemistry and engineering programs are placed in the course catalog. His explanation is based on data from the universities of Oxford and Sheffield where enrollment is 20-30 students a year and the green courses are listed in chemistry and chemical engineering curricula instead of being listed under a separate green curriculum in the course catalog. Howdle also thought that even differences in language, such as "octylamine" versus "octyl amine," between the United Kingdom and the United States create hurdles for the green community because they create cultural differences.

Vanasupa shared Cal Poly's materials engineering undergraduate program marketing experiment with the workshop participants. One part of the marketing experiment was changing the e-mail announcement about the program from an "Extreme Action" theme to a more subtle theme: trees with music. A second part of the marketing experiment was redesigning the Web pages with more photographs depicting a diverse array of people (see Figure 5.1). Motivations for the changes were to have students practice reflection and to create an image that the field embraces diversity.[1] Cal Poly also wants to understand the issues driving humanistic students.

The number of U.S. citizens versus international students in graduate science and engineering programs is another issue. Despite a 15-year investment in science and engineering education, there is a decreasing trend in the number of engineering and chemistry degrees. For example, in 2000 approximately 5,000 Ph.D. degrees in the physical sciences were granted in the United States compared with approximately 25,000 Ph.D. degrees granted in China.[2]

Communication through advertising, teaching, and educational materials are also seen as useful to get the message out and market green chemistry. The many American Chemical Society (ACS) publications Kathryn Parent presented are one avenue. Both the ACS's *Journal of Chemical Education* and *The CPT* (Committee on Professional Training) news-

[1]http://mate.calpoly.edu/quest.

[2]National Science Board. 2002. *Science and Engineering Indicators 2002*. Arlington, VA: National Science Foundation.

FIGURE 5.1 Web shot of California Polytechnic's Materials Engineering Web site. SOURCE: Vanasupa, L. Where Do We Go from Here? Addressing the Human Dimension of Curricular Design. Presentation at the National Academies Chemical Sciences Roundtable Green Chemistry and Engineering Education Workshop. November 8, 2005.

letter, which Fleming Crim spotlighted, are excellent forums. In addition to historical media outlets external to a school, such as radio and newspaper, working with a school's radio station or journalism school, are other broadcasting avenues for all levels of education. In addition, collaborations on energy initiatives that are springing up on many campuses and in many industries (not just the petrochemical industry) are also opportunities for communication between groups who do not routinely communicate with one another, such as civil engineers and energy economists.

ADDITIONAL GREEN CURRICULAR IDEAS

Several ideas for curricular development not mentioned in existing green material or in the process of development were noted. Both David Shonnard from Michigan Technological University and Stanley Manahan from the University of Missouri brought up the idea of industrial ecology or the science of sustainability. According to Shonnard, industrial ecology is an interdisciplinary framework for designing and operating industrial systems as living systems interdependent with natural systems. Therefore, there is a balance between environmental and economic performance with local and global ecological constraints. Industrial ecology comprises several tools and systems, such as Life Cycle Assessments (LCA), according to David Allen from the University of Texas-Austin. A standard definition of an LCA is an objective process to evaluate the environmental burdens associated with a product, process, or activity by identifying energy, materials, and wastes in order to evaluate and implement opportunities to affect environmental improvements. Material and energy flow analyses (e.g., mass balancing) for a variety of scales, such as an individual business, industrial sector, or an entire economy, that are measuring environmental performance are part of LCAs.

Legislative issues in occupational and public health seem to be beyond what is considered standard for life-cycle assessments and completely unrelated to green chemistry and engineering; however, there are important connections that should be noted. Mike Wilson from University of California, Berkeley, School of Public Health cited the issue of work-related exposures in the United States, in California in particular, as examples of green occupational health issues. Work related hazardous exposures represent 60,000 deaths in the United States every year, 7,000 in California alone, and is therefore a major public health issue. Specific examples of green-chemistry-related occupational and public health legislative issues include (1) the phaseout of perchloroethylene and other chlorinated solvents from vehicle repair; hexane was substituted for perchlorethylene in California and across the United States with deleterious effects; and (2) bromopropane has been introduced as a substitute for chlorofluorcarbons. Julie Zimmerman from the Environmental Protection Agency and the University of Virginia added that environmental and human health impacts are typically viewed as an outcome from LCA rather than an integrated part of life-cycle management. Therefore it is important to introduce engineers and chemists, most of whom probably do not know what lethal dose 50[3] is, to environmental health and biomedical topics such as toxicity, toxicology, and epidemiology.

A second curriculum topic that Eric Beckman from the University of Pittsburgh proposed is chemical product design with sustainability being the design goal. Product design classes are common in other types of engineering, such as mechanical and electrical, but have eluded chemical engineers. Of the 6.5 billion people on the planet, most do not know or have any background in chemical product design. A product design class would be a multidisciplinary subject, so ideally there would be chemists and business students in addition to chemical engineers taking the class. In addition, teaching the design philosophy that green products are almost as cheap and almost as good is not good enough. The products need to be cheaper, reliable, and green.

In the area of curriculum, access to information and resources is seen as one of the biggest challenges, which goes back to the topic of marketing. According to Howdle, Parent, and Haack, one simple issue is just getting people in touch with the resources. Jorge Vanegas from Texas A&M University, formerly of Georgia Institute of Technology, stressed that creating common classes for chemistry, chemical engineering, and other students within the standard engineering and chemistry curricula is a solution versus teaching in a silo, single-channel fashion. Crim from the University of Wisconsin sees more materials being freely available, as well as custom publishing, so that the cost of materials and books decreases because custom publishing could create iTunes™ like databases of laboratory experiments. People will be able to selectively purchase only the experiments they want.

[3]Lethal dose 50 (LD50) is the dose at which 50 percent of an exposed animal population dies.

RESEARCH EXTENSIVE UNIVERSITIES

Although there are signs that research extensive universities (R1s) are becoming more involved with green education in direct and indirect ways, the lack of presence and support of R1 universities for green chemistry and green engineering education was a theme brought up several times during the workshop. Parent sees R1 involvement as low. Richard Wool from the University of Delaware sees R1s 10 years behind in green education. Howdle strongly articulated that R1s are conflicted between having the traditional R1 attitude of research powerhouses, and still wanting to install more fume hoods to work with even more hazardous materials versus the integration of greenness that would dilute the skills of the graduate students and postdocs. Kenneth Doxsee from the University of Oregon believes that since R1s are major feeders for industrial employers and next generations of faculty, they have a responsibility to embrace green principles. Doxsee thought that the presence of faculty from MIT, Cornell, and Wisconsin at this Chemical Science Roundtable workshop was a sign that R1s are seeing value in green education.

There was both agreement and disagreement that the trickle-down theory of green research done by faculty, postdocs, and graduate students influences the undergraduate curriculum in a positive way. Three examples of the many intellectual opportunities green research gives researchers, according to Crim, are Tyler McQuade (Cornell University) researching telescoping of reactions, Barry Trost (Stanford University) emphasizing atom economy, and Shannon Stall (University of Wisconsin) focusing on inorganic catalysis. Crim explained that translating this research into education critical because just "cherry picking" reactions that can be made to look very green can turn out to be very distant from what people are doing in R1 universities. The Grignard reaction is a good example of this. The Grignard reaction is still traditionally taught in chemistry classes, but metal-catalyzed coupling reactions, not Grignards, are more commonly used in the laboratory. Crim suggests that the problem is how effective the R1 faculty members are in translating the green research into teaching, with some professors being more effective than others. In addition, Crim believes that the R1 universities could use the green textbooks and material that undergraduate colleges use to prevent "reinventing the wheel" and ease the overload burden.

BUSINESS EDUCATION

A common theme throughout the workshop was incorporating green ideas into business education. According to Warner, training people just to work for corporations is not enough anymore, and an emphasis on entrepreneurship is needed. Two examples of business education activities are Pat Hogan's business club at Suffolk University and Tyler McQuade's efforts to develop relationships between Cornell's chemistry department and its business school. The fact that green business efforts for undergraduate, graduate, and faculty are in motion is evidence that many see it as important. Parent sees one of the issues of selling green education to business people is making it clear how they will gain from using green principles. Parent notes, however, that it is the scientists and engineers pushing the efforts, not the business community and it is therefore currently a one-sided push.

GREEN ETHICS

Green ethics, or the social responsibility to improve the environment, was another curriculum item that came up in discussion several times. In addition to many of the attendees who deem green ethics important, it is also information the students want. Vanasupa said that many students are very attracted to the idea of "making a difference" and "service to humanity." According to Vanasupa, ethics have been a peripheral subject in many schools and having the material as an integral part of the curriculum is a goal with a variety of solutions available.

Cliff Davidson from Carnegie Mellon University posed the question of "who will teach the ethics?" Davidson suggested that expert ethicists and humanities professors were seen as appropriate people to teach as well as help develop the curriculum. Vanasupa explained that because faculty members are often already burdened with a full workload, one solution has been to outsource to the experts. At the same time, she said that the outsourcing can also create a disconnect between the faculty and material covered. One solution to the disconnect is coteaching, although there can initially be problems with the administration and infrastructure.

CONCLUSION

By the close of the two days, the attendees of this Chemical Sciences Roundtable workshop had covered a wide array of green chemistry and green engineering education efforts and ideas for all levels of education. The existing and developing efforts at the pre-college, undergraduate, graduate, faculty, and industry levels discussed cover many formats:

- Workshops;
- Videos;
- Computer programs;
- Research programs;
- Degree programs;
- Booklets;
- Textbooks;
- Competitions;
- Websites;
- Databases; and
- Distance education.

Developing curricular ideas around the issues of marketing, occupational health, business education, R1s, and green ethics are also seen as important for the future of green chemistry and green engineering education. The workshop served as a forum to organize a core of leaders who hope to further facilitate, catalyze, and integrate green chemistry, engineering, and policy into historical curricula.

Comparing the ideas about green chemistry and engineering education that participants identified in the pre-workshop survey with what the attendees were able to discuss and rally around indicates consistency of trends. In the pre-workshop survey the majority of the respondents (76 percent) felt an integrated approach for teaching the material was more effective than teaching separately. A similar idea, presented in the overarching marketing section, came out repeatedly during the two-day workshop. In the area of impediments to incorporation, the respondents did not identify one factor as dominant. Instead books, lecture materials, colleague resistance or lack of awareness, and a crowded curriculum were each about equally important (about 20 percent each). The lack of materials and crowded curriculum, as well as the lack of awareness of materials, mentioned in the overarching section mirrored the survey results. Prior to the workshop, the attendees indicated that green education was best targeted at all undergraduate levels (67 percent), as well as at the freshmen level (17 percent). The presentation of so many efforts at all levels of education during the workshop indicates that there is an interest for some kind of education at all levels. In addition, the particular breakout group, Green Chemistry and Engineering in Future Curricula, felt that a specific degree program is best targeted at the graduate level since undergraduates are trained to be generalists. During the workshop the attendees' discussion of the benefits of green education agreed with the survey's findings: (1) enthusiasm (35 percent); (2) recruitment and retention (23 percent); and (3) increased job opportunities (18 percent). Overall, the attendees also agreed with the survey that green teaching aids the teaching of historical curricula (100 percent) and acts as a multidisciplinary tool (94 percent). In addition, the workshop discussion identified savings in laboratory equipment, chemicals, and supplies as huge benefits. Although the survey clearly indicated that the attendees felt that a lack of funding (91 percent) was an issue and during the workshop funding was occasionally mentioned, the workshop focused on the many content, growth, and implementation ideas.

Appendix A

Summary of Pre-Workshop Participant Survey

Forty-three workshop participants answered a 10-question survey to gather information on the details of Green Chemistry (GC) and Green Engineering (GE) education issues of interest to the attendees. The mix of multiple-choice, yes-no, and open-ended questions cover who is interested, how should it be taught, who benefits, and funding. The questions together with the tabulated answers are listed below.

Question #1	Academe	Industry	Government	Nonprofit	Other	Integrated
Participants sector	74%	3%	11%	6%	3%	3%

Question #2	Integrated	Separate	Both
GC/GE Integrated or separate course	76%	9%	15%

Question #3	Books	Lecture Materials	Colleague Resistance/Lack of Awareness	Crowded Curriculum	Institutional Resistance	Other
Impediments to incorporation	16%	20%	23%	22%	9%	10%

Question #4	Freshmen	Integrated	Upper-Level Undergraduate	Graduate Level	Other
At what grade level	17%	67%	8%	0%	8%

Question #5	Enthusiasm	Recruitment & Retention	Increased Job Opportunities	Other
Largest benefit of GC/GE education to student	35%	23%	18%	24%

Question #6	Yes	No	Some	Unsure
Sufficient funding/support for GC/GE education	3%	91%	3%	3%

Question #7	Yes	No
GC/GE education assist in teaching traditional technical concepts	100%	0%

Question #8	Yes	No
GC/GE helpful teaching multidisciplinary	94%	6%

Question #9	Top 5 Open-Ended Responses
What is the single most important action that would help advance the implementation of green chemistry and green engineering education?	Funding for more research, curriculum development, teaching materials, U.S. chemical policy reviews, and other GC/GE causes
	Educational materials and textbooks
	Awareness at all levels of education. professional societies, and industry
	Employer demand
	Required curriculum in classroom

Question #10	Top 5 Responses
Who is responsible for taking action?	Federal, state, and local government
	Educational institutions
	Industry, especially those involved with GC/GE
	Professional societies (e.g., American Chemical Society)
	All of the above

Appendix B

Summary of Green Chemistry and Green Engineering Education Efforts

PRE-COLLEGE GC/GE EDUCATION EFFORTS

Pre-College Level	Educational Tool	Organization	Contact
Elementary and above	Textbook: *Chemistry Experimentation for All Ages*		Ken Doxsee
Elementary/Middle School	Pollution workbooks & programs: Pollution Solutions Pollution Prevention: A Story of Carbopond Cleanup	Suffolk University/Rohm & Haas	Pat Hogan
Middle School	Energy and Pollution Prevention Program	Michigan Technological University	James Mihelcic
Middle School	Green Chemistry and Environmental Sustainability: A Middle School Module	Pfizer	Berkeley Cue
High School	*Chemistry in the Community* *Introduction to Green Chemistry*	American Chemical Society	Kathryn Parent
High School	Distance Education Green Chemistry Course	University of Oregon/Worcester State College	Julie Haack/ Margaret Kerr

UNDERGRADUATE GC/GE EDUCATION EFFORTS

Undergraduate Level	Educational Tool	Organization	Contact
Undergraduate	Green Chemical Engineering Material Framework	University of Texas, Austin	David Allen
Sophomore Level	Fall & Winter Term Green Laboratory Experiments Organic Chemistry Laboratory	University of Oregon	James Hutchison
Interdisciplinary Multilevel	Environment Across the Curriculum Multidisciplinary Programs	Carnegie Mellon University	Cliff Davidson
Nonchemistry Major	*Chemistry for Changing Times*, 10th ed., Chemistry Textbook for Nonmajors		John W. Hill, Doris K. Kolb

Graduate Level	Educational Tool	Organization	Contact
Multilevel Undergraduate Courses	Green Chemistry for Process Engineering	University of Nottingham	Steven Howdle
Advance Chemistry Undergraduate Course	Industrial and Applied Green Chemistry	York University	John Andraos
Undergraduate	Textbooks: *Chemistry in Context* *Real-world Cases in Green Chemistry* *Greener Approaches to Undergraduate Laboratory Experiments*	American Chemical Society	Kathryn Parent
Undergraduate	Video: Green Chemistry – Innovations for a Cleaner World Companion	American Chemical Society	Kathryn Parent
Undergraduate	Textbook: *Environmental Chemistry*, 3rd Edition		Michael Cann
Senior Undergraduate Class	Green Engineering – Out of this World CHEG 667 – Senior Undergraduate Class	University of Delaware	Richard Wool
Senior/Grad students	National P3 Design Competition: People, Prosperity, and the Planet	EPA	Julie Zimmerman
Undergraduate	Green-Soil & Water Analysis at Toad Suck (GSWAT) Laboratory Course	Hendrix College	Liz Gron
Undergraduate Honors	Green Business Seminar	Suffolk University	Pat Hogan
Undergraduate	Green Engineering Freshmen Curriculum Module	Rowan University	David Shonnard
Undergraduate	NSF Grant – Sustainability Principles for Curriculum	Cal Poly, San Luis Obispo	Linda Vanasupa

GRADUATE GC/GE EDUCATION EFFORTS

Graduate Level	Educational Tool	Organization	Contact
Graduate	Ph.D. Chemistry Program	University Massachusetts Lowell	John Warner, Amy Cannon
Graduate, Postdoctoral	Sustainable Chemistry in the Pharmaceutical Industry Green Chemistry Workshop	Pfizer – Groton Labs	Berkeley Cue
Graduate, Postdoctoral	Green Chemistry & Business Case Studies, including contact with Campus Business Clubs	Cornell University	Tyler McQuade
Graduate, Postdoctoral	Partnership Program with Southern University Science Education Program with Peace Corps	Michigan Technological University	James Mihelcic

FACULTY GC/GE EDUCATION EFFORTS

Faculty Level	Educational Tool	Organization	Contact
Faculty	Sophomore Green Organic Laboratory Education & Training	University of Oregon	James Hutchison

Faculty	Green Chemistry Educational Ambassador Sites	University of Oregon	Julie Haack
Faculty	International Faculty Training	University of Oregon	Ken Doxsee
Faculty	Textbook: *Going Green – Integrating Green Chemistry into Curriculum*	American Chemical Society	Kathryn Parent
Faculty	Center for Sustainable Engineering (CSE) Workshop: Faculty training to develop and sustain environmental programs; Bookbuild Web site	Carnegie Mellon University	Cliff Davidson
Faculty	ASEE Green Engineering Web site	Rowan University	David Shonnard
Faculty	DICE Recruitment Program	Nottingham	Steve Howdle

INDUSTRY EDUCATION EFFORTS

Industry	Educational Tool	Organization	Contact
Internal (In-house)	Pfizer Green Chemistry Achievements	Pfizer	Berkeley Cue
Internal (In-house)	Process Position– Green link between Research & Manufacturing	Merck	John Leazer

GENERAL GC/GE EDUCATION EFFORTS

General	Educational Tool	Organization	Contact
Educational Materials	Frontiers in Chemical Engineering Education	MIT	Jeffrey Steinfeld
Chemistry Educational Material	GEMS – Green Education Materials for Chemists	University of Oregon	Julie Haack
Textbook	*Green Engineering: Environmentally Conscious Design of Chemical Processes*	Michigan Technological University	David Shonnard/David Allen
Business School Education	Business Case studies, Chemistry & Business School Collaborations	Green Chemistry Institute	Kathryn Parent
Electronic Tools	Green Chemistry Alternative Selection Protocol NEMI Analytical Method Database	Green Chemistry Institute	Kathryn Parent
Web site	Joseph Breen Chemistry Awards	Green Chemistry Institute	Paul Anastas
Web site	Canadian Green Chemistry Network		John Andraos
DVD	Meeting Global Challenges	American Chemical Society	Kathryn Parent
Textbook	*Bio-based Polymers & Composites*	Academic Press	Richard Wool
Textbook	*Green Chemistry and the Ten Commandments of Sustainability*	ChemChar Research Inc.	Stanley E. Manahan
Book	*Sustainability in the Chemical Industry: Grand Challenges and Research Needs*	National Research Council	National Academies Board on Chemical Sciences and Technology

Appendix C

Workshop Agenda

MONDAY, NOVEMBER 7, 2005

Opening

8:30-9:00	Paul Anastas, ACS Green Chemistry Institute

Current Status

9:00-9:30	David Allen, University of Texas, Austin
9:30-10:00	James Hutchison, University of Oregon
10:00-10:30	Steven Howdle, University of Nottingham (on sabbatical at University of Pittsburgh)

Current Status Panel

10:45-11:00	Amy Cannon, Rohm and Haas
11:00-11:15	Berkeley Cue, Pfizer (retired)
11:15-11:30	Kenneth Doxsee, National Science Foundation (University of Oregon)
11:30-11:45	Tyler McQuade, Cornell University
11:45-12:15	Panel Discussion

Tools and Materials

1:15-1:45	Julie Haack, University of Oregon
1:45-2:15	David Shonnard, Michigan Technological University
2:15-2:45	John Andraos, York University (Canada)
2:45-3:15	John Warner, University of Massachusetts, Lowell

Tools and Materials Panel

3:30-3:45	Michael Cann, University of Scranton
3:45-4:00	Eric Beckman, University of Pittsburgh
4:00-4:15	Kathryn Parent, Green Chemistry Institute (ACS)
4:15-4:30	Richard Wool, University of Delaware
4:30-5:00	Panel Discussion

TUESDAY, NOVEMBER 8, 2005

Where Do We Go from Here?

8:30-9:00	F. Fleming Crim, University of Wisconsin
9:00-9:30	Cliff Davidson, Carnegie Mellon University
9:30-10:00	Terrence Collins, Carnegie Mellon University
10:00-10:30	Julie Zimmerman, University of Virginia (EPA)
10:30-10:45	Break

Where Do We Go from Here? Panel

10:45-11:00	Linda Vanasupa, California Polytechnic State University
11:00-11:15	John Leazer, Merck & Company
11:15-11:30	James Mihelcic, Michigan Technological University
11:30-11:45	Jorge Vanegas, Georgia Institute of Technology
11:45-12:15	Liz Gron, Hendrix College
12:15-12:45	Panel Discussion

Introduction of Breakout Sessions

1:30-1:45 Frankie Wood-Black, Conoco-Phillips Company

Breakout Sessions (concurrent)

1:45-3:15
1. Green Chemistry & Green Engineering in Future Curricula
2. What Materials, Programs & Tools are Needed
3. What is Needed to Achieve Interdisciplinary Approaches?

Breakout Sessions (concurrent)

3:30-5:00
1. Green Chemistry & Green Engineering Industry & Education
2. MGreen Chemistry, Green Engineering and the New Faculty
3. Creating Incentives, Removing Impediments

5:00 Meeting Adjourns

Appendix D

Biographies

ORGANIZERS

Paul Anastas is the director of the Green Chemistry Institute in Washington, DC. Until June of 2004 he served as assistant director for environment at the White House Office of Science and Technology Policy where his responsibilities included a wide range of environmental science issues, including furthering international public-private cooperation in areas of Science for Sustainability (such as Green Chemistry). Prior to coming to OSTP in October 1999, Dr. Anastas served as the chief of the Industrial Chemistry Branch of the U.S. Environmental Protection Agency since 1989. During that period he was responsible for regulatory review of industrial chemicals under the Toxic Substances Control Act and the development of rules, policy, and guidance. In 1991 he established the industry-government-university partnership Green Chemistry Program, which was expanded to include basic research, and the Presidential Green Chemistry Challenge Awards. Prior to joining the EPA, he worked as an industrial consultant to the chemical industry in the development of analytical and synthetic chemical methodologies. Dr. Anastas received his M.A. and Ph.D. in organic chemistry from Brandeis University and his B.S. in chemistry from the University of Massachusetts at Boston.

Frankie Wood-Black is the director, Business Services for Downstream Technology, Conoco-Phillips with responsibility for finance, business analysis, training, and assets for Downstream Technology. Prior to this, Frankie was the technology services marketing manager for Phillips and was responsible for in-sourcing research and development activities into the Bartlesville Technical Center and also served as quality assurance team leader at the Borger Refinery and Natural Gas Liquids Center. Wood-Black began her career with Conoco-Phillips in 1989 in Bartlesville, Oklahoma, as a research scientist for research and development. In 1994 she was transferred to the Woods Cross Refinery as an environmental scientist whose job responsibilities included regulatory compliance for air, community-right-to-know, and the Toxic Substance Control Act. She was transferred in 1998 to Corporate Health, Environment, and Safety in the Property Risk Management Group to become site manager for nonoperating sites before her relocation to Borger, Texas, in 1999. She received a Bachelor of Science degree in physics with a minor in chemistry from Central State University (now the University of Central Oklahoma) in Edmond, OK, in 1984. She attended Oklahoma State University and received a doctorate in physics in 1989 and completed her M.B.A. in December 2002. Wood-Black has been active in numerous professional activities and serves as the Conoco-Phillips representative on Corporation Associates of the American Chemical Society. She is a contributing editor of the *Journal for Chemical Health and Safety* with her coauthored column "CHAS Netways." She has one patent, ten technical publications, and has coauthored a book entitled *Emergency Preparedness Planning—A primer for Chemists.* Wood-Black regularly makes presentations at the American Chemical Society National meetings. She is a registered environmental manager.

SPEAKERS

David Allen is the Melvin H. Gertz Regents Chair in Chemical Engineering and the director of the Center for Energy and Environmental Resources at the University of Texas at Austin. His research interests lie in air quality and pollution prevention. He is the author of four books and over 150 papers in these areas. The quality of his research has been recognized by the National Science Foundation (through the Presidential Young Investigator Award), the AT&T Foundation (through an Industrial Ecology Fellowship), the American Institute of Chemical Engineers (through the Cecil Award for contributions to environmental engineering), and the State of Texas (through the Governor's Environmental Excellence Award). In addition, Dr. Allen is actively involved in developing green engineering educational materials for the chemical engineering curriculum. His most recent effort is a textbook on design of chemical processes and products, jointly developed with the EPA. Dr. Allen received his B.S. degree in chemical engineering, with distinction, from Cornell University in 1979. His M.S. and Ph.D. degrees in chemical engineering were awarded by the California Institute of Technology in 1981 and 1983. He has held regular faculty appointments at UCLA and the University of Texas and visiting appointments at the California Institute of Technology and the University of California, Santa Barbara; he joined the University of Texas in 1995.

John Andraos earned a Ph.D. in 1992 from the University of Toronto, studying the ketene hydration reaction by flash photolysis. He set up conventional (microsecond) and nanosecond laser flash photolysis apparatuses for the Reaction Intermediates Group at Toronto. He then did postdoctoral work at the University of Ottawa, where he discovered ketene zwitterion intermediates, and at the University of Queensland, Australia, where he discovered the first example of a 1,3-sigmatropic rearrangement in acylketenes by Nuclear Magnetic Resonance Spectroscopy. He also developed computational protocols for the evaluation of kinetic data obtained in heterogeneous media, such as zeolites and low-temperature argon matrices. Since his appointment as lecturer and course director at York University in 1999, he has taught and developed courses in organic chemistry. In 2002 he launched the first industrial and green chemistry course in the history of the Department of Chemistry at York. He has published 35 research articles in refereed journals, 11 of these as an independent researcher. He has given invited addresses to Concordia University, the University of Western Ontario, and the University of Toronto. In 2000 he launched the CareerChem Web site, which is an in-depth resource for tracking and cataloguing all named things in chemistry and physics, chronicling the development of chemistry through scientific genealogies, and supplying career information to young researchers and students for placement in academic and industrial positions worldwide. Since 2000 he has given career workshops at the annual Canadian Society of Chemistry conference for students and postdoctoral fellows. His awards of recognition include the Junior Research Award from the Australian Research Council in 1996, and he is currently the president of the University of Toronto Sigma Xi Chapter.

Michael Cann was born and raised in the Saratoga region of upstate New York and attended Marist College, where he earned his B.A. in chemistry in 1969. Mike received his M.A. and Ph.D. in organic chemistry from State University of New York, Stony Brook, in 1972 and 1973. He was a postdoctoral fellow at the University of Utah (1973-1974), and a lecturer at the University of Colorado-Denver (1974-1975). Since 1975 he has been a faculty member at the University of Scranton. He is also the codirector of the environmental science program. His areas of interest encompass nitrenium ions, nitrogen heterocycles, and green chemistry. His interests in green chemistry consist of microwave-assisted organic reactions, solvent-free organic reactions, and green chemistry education.

Amy Cannon recently graduated as the first Ph.D. in green chemistry at the University of Massachusetts, Boston, working in Professor John Warner's research group. Her research at UMass Boston involved the environmentally benign synthesis of photoactive materials, including titanium dioxide semiconductors, photoresist polymers, and novel spiropyran photoactive materials. Amy received her B.S. in chemistry from Saint Anselm College in Manchester, NH, and worked for the Gillette Company as an analytical chemist for five years before returning to graduate school. She was awarded the Kenneth G. Hancock Memorial Award in Green Chemistry in 2004 for her work on titanium dioxide semiconductors and their application in dye-sensitized solar cells. Amy currently works as a chemist for Rohm & Haas Electronic Materials in Marlborough, MA, where she is developing silicon polymeric materials for optical electronic devices. She is also an adjunct professor of green chemistry at the University of Massachusetts, Lowell.

Terry Collins is the Thomas Lord Professor of Chemistry at Carnegie Mellon University, where he directs the Institute for Green Oxidation Chemistry. He is also an honorary professor at the University of Auckland, New Zealand. Professor Collins earned his B.Sc. (1974), M.Sc. (1975), and Ph.D. (1978) degrees from the University of Auckland, where his graduate advisor was Warren R. Roper. After postdoctoral work at Stanford University with Jim Collman, he joined the faculty of Caltech in 1980 and the faculty of Carnegie Mellon University in 1987. He has a number of research awards, including the 1998 Presidential Green Chemistry Challenge Award. Professor Collins writes and lectures widely on the possibilities before chemists to develop vibrant new economies to promote sustainability. His research program is focused on greening oxidation technologies by designing nontoxic catalysts for activating the natural oxidants, hydrogen peroxide and oxygen, for nonpolluting oxidations. Professor Collins serves on the editorial advisory boards of *C&E News* and *Environmental Chemistry*. He is involved in steering or contributing to numerous international conferences and educational programs aimed at promoting green chemistry.

F. Fleming Crim is the John E. Willard and Hilldale Professor of Chemistry at the University of Wisconsin-Madison.

He received his Ph.D. from Cornell University in 1974 and worked on semiconductor manufacturing techniques at the Engineering Research Center of Western Electric Co. until 1976. He then spent a year as a director's postdoctoral staff member at Los Alamos National Laboratory and moved to Madison as an assistant professor in 1977. He was chair of the department from 1995 to 1998, and is currently chair of the Committee on Professional Training of the American Chemical Society. His research in chemical reaction dynamics uses lasers to explore and control the course of chemical reactions in both gases and liquids. He is a member of the National Academy of Sciences.

Berkeley W. Cue, known as Buzz to most of us, consults with several technology companies who serve the pharmaceutical industry to create innovative solutions for pharmaceutical science and manufacturing challenges. Most recently, he was responsible for pharmaceutical sciences at Pfizer's Groton, Connecticut, R&D site. He created and led Pfizer's green chemistry initiative. Dr. Cue retired from Pfizer in April 2004 after almost 29 years, but he continues his mission of advancing green chemistry in the pharmaceutical industry. Since he retired in 2004 he has given more than a three dozen presentations on various aspects of green chemistry in the pharmaceutical industry. He received a B.A. with honors from the University of Massachusetts, Boston (1969); his Ph.D. (organic chemistry) from the University of Alabama (1973); completed postdoctoral research at the Ohio State University (1974); and was a National Cancer Institute Research fellow at the University of Minnesota (1975). He is a member of the Gamma Sigma Epsilon chemistry honors fraternity at the University of Alabama. In 2003 he received the Pfizer Groton Labs Green Chemistry Award and was presented the Seldon Award by University of Massachusetts, Lowell, for his contributions to green chemistry.

Liz Gron is an associate professor of chemistry at Hendrix College. Her research interests focus on green chemistry, specifically in the area of organic reactions in near-critical water. The research goal is to replace nonrenewable petrochemical solvents while exploiting the unique properties of extremely hot water to investigate underlying mechanistic interactions between the reactants and the solvent. Additionally, Professor Gron has developed educational materials for the introductory chemistry curriculum. At present, this work focuses on designing green experiments that teach analytical and environmental chemistry to introductory chemistry students. She has taught courses in general, analytical, and inorganic chemistry while at Hendrix. Liz Gron earned her B.A. in chemistry at Colgate University and her Ph.D. in inorganic chemistry with Arthur B. Ellis at the University of Wisconsin-Madison, 1987. She was a postdoctoral fellow and an industrial research liaison at the Department of Chemical Engineering, University of Delaware, before starting at Hendrix in 1994. During her latest sabbatical, Professor Gron held a position of visiting assistant professor of chemical engineering, Massachusetts Institute of Technology, while working as a visiting scientist with Jefferson Tester. Professor Gron is an active member of Project Kaleidoscope's faculty for the 21st century and has presented on aspects of chemistry curriculum at a number of related workshops.

Julie A. Haack is a senior instructor and assistant department head for chemistry at the University of Oregon, where her work has focused on the incorporation of green chemistry principles into the introductory chemistry curriculum for both science and nonscience majors. She is a leader in facilitating the identification, development, and dissemination of green chemistry educational materials throughout the chemistry curriculum.

Dr. Haack has developed Greener Education Materials (GEMs)s for chemists, a database of green chemistry laboratory exercises and educational materials that is enabling educators at all levels to easily identify and incorporate green chemistry into their curriculum. GEMs is also becoming instrumental in supporting the development of a growing community of green chemistry educators interested in new materials development. Dr. Haack received her Ph.D. in biology from the University of Utah (1991) and her B.S. in chemistry from the University of Oregon (1986). She completed postdoctoral work in pharmacology at the University of North Carolina at Chapel Hill and in biophysics as a Howard Hughes research associate at the University of Oregon.

Steve Howdle holds a chair of chemistry at the School of Chemistry, University of Nottingham, and prior to this held a distinguished Royal Society University Research Fellowship (1991-1999). In 2001 he was a recipient of both the Jerwood-Salters Environment Award and the Corday-Morgan Medal and Award of the Royal Society of Chemistry. In 2003 he received a Royal Society–Wolfson Research Merit Award. Steve's academic interests focus on the utilization of supercritical carbon dioxide for polymer synthesis, polymer processing and preparation of novel polymeric materials for tissue engineering and drug delivery. A more detailed description of his research can be viewed at http://www.nottingham.ac.uk/~pczctg/Index.htm. He has to his credit over 180 academic papers, reviews, and patents, and is also the driving force behind a spin-off company, Critical Pharmaceuticals Ltd., which was founded upon his academic work. Steve obtained a B.Sc. in chemistry from the Victoria University of Manchester in 1986 and a Ph.D. on "Spectroscopy in Liquefied Noble Gases" from the University of Nottingham in 1989.

James ("Jim") Hutchison is currently professor of chemistry and director of the Materials Science Institute at the University of Oregon (UO). Since joining the faculty at UO in the fall of 1994, Hutchison and his research group have worked to design and make new functional molecules, materials, and nanomaterials. His specific research interests are preparation and study of nanoscale materials, surface, and polymers for applications such as nanoelectronics, biocompatibility, and environmental remediation. He has pioneered the emerging field of green (environmentally friendly) nanoscience. He is a leading chemical educator, having played key roles in developing the UO's nation-leading program in green organic chemistry and designing the Materials Science Institute Graduate Internship Program in Semiconductor Processing.

Before joining the faculty at the University of Oregon, Hutchison was a National Science Foundation postdoctoral fellow with Professor Royce W. Murray at the University of North Carolina at Chapel Hill, where studied the surface and electrochemistry of monolayers on gold films and nanoparticles. He received his Ph.D. in 1991 at Stanford University under the direction of Dr. James P. Collman, studying the binding and redox chemistry of hydrogen, oxygen, and dinitrogen by specifically designed cofacial diporphyrin catalysts. His undergraduate degree, a B.S. in chemistry, was completed in 1986 at the University of Oregon.

D. Tyler McQuade, assistant professor of chemistry and chemical biology at Cornell University, began his faculty position in 2001. He is currently a Dreyfus, 3M, Rohm and Haas, Beckman, and NYSTAR Young Investigator and one of the 2004 MIT Tech Review 100. McQuade was born in Atlanta, GA, and raised in the Santa Cruz mountains of California. He received a B.S. in chemistry and a B.S. in biology from the University of California, Irvine, and a Ph.D. in chemistry from University of Wisconsin-Madison under the guidance of Professor Samuel Gellman. His education was completed by a NIH Fellowship at MIT with Professor Timothy Swager. The McQuade group is focused on creating synthetic systems that use site isolation (via encapsulation) and selectivity (via recognition) to carry out multistep syntheses with greater efficiency. The group's multidisciplinary environment, using tools from biology, chemistry, and materials science, is yielding both polymers and small molecules that will provide the building blocks for the next generation of reagents for sustainable process chemistry. Many of these early innovations nucleated the Sustainable Pharmaceutics enterprise that won Cornell's Big Red Ventures 2005 *Business Idea Competition*.

James R. Mihelcic is a professor of civil and environmental engineering at Michigan Technological University (Houghton, MI) and an adjunct graduate faculty member at Southern University and A&M College (Baton Rouge, LA). He also serves as the codirector of the Sustainable Futures Institute (www.sfi.mtu.edu) and director of the Master's International Program in Civil and Environmental Engineering (www.cee.mtu.edu/peacecorps). This latter program allows graduate students to work on sustainable development issues overseas while combining graduate studies with engineering service in the U.S. Peace Corps. His teaching and research interests are in the areas of sustainability and green engineering; biological and chemical processes; and water and sanitation issues in the developing world. He has also conducted extensive research in developing methods to estimate environmental properties of chemicals based on chemical structure and integrated these estimation methods with models of environment risk, the economy, and environmental fate and transport to evaluate emerging chemicals and sustainable economic activities.

Kathryn E. Parent is a staff associate for the Green Chemistry Institute at the American Chemical Society. Parent leads educational initiatives at GCI. She received her B.S. in chemistry from George Fox University in Newberg, Oregon. Following graduation she worked as a research assistant at Oregon Health Sciences University. Parent became involved with green chemistry in 2000 when she became a graduate teaching fellow in the Chemistry Department at the University of Oregon. She moved to Washington, DC, to begin working for GCI in 2002. Her major interests include green chemistry, education, and community outreach. Her recent efforts include editor for *Going Green: Integrating Green Chemistry into the Curriculum* (2004); conference administrator for the joint meeting of the Second International Conference on Green and Sustainable Chemistry and the Ninth Annual Green Chemistry and Engineering Conference (2005); and program director for the ACS Summer School on Green Chemistry (2005). She is a frequent contributor to the ACS magazine for student affiliates, *Chemistry Magazine*.

David R. Shonnard received a Ph.D. in chemical engineering from the University of California, Davis, in 1991 and conducted postdoctoral research at Lawrence Livermore National Laboratory (1990-1992). His research and teaching interests include green engineering, biotechnology and bioprocessing, enzyme engineering, and environmental life-cycle impact assessment. At MTU he is a research and education pioneer in areas of environmental impact assessment for chemical products and processes and coauthor of the textbook *Green Engineering: Environmentally Conscious Design of Chemical Processes*, Prentice Hall, 2002. He has consulted with several major chemical manufacturers (BASF, UOP) on the use of life-cycle assessment to evaluate and improve environmental performance. He is the recipient of the 1998 NSF/Lucent Technologies Foundation Industrial Ecology Research Fellowship and the 2003 Ray Fahien Award of the American Society for Engineering Education. He has received over $2 million in research funding at Michigan Technological University and published over 50 peer-reviewed research papers.

Linda Vanasupa is currently serving as chair of California

Polytechnic State University's Materials Engineering Department (mate.calpoly.edu) and associate director of Cal Poly's Center for Sustainability in Engineering (csine.calpoly.edu). As principal investigator of a National Science Foundation Department-Level Reform grant, she and her colleagues are exploring the design of engineering learning experiences that promote systems thinking, responsible global citizenship, and retention of underrepresented individuals. Her recent research interests have included thin film processing and characterization and materials degradation in hydrogen fuel cells. Professor Vanasupa has degrees in material science and engineering (Ph.D. Stanford University, 1991; M.S. Stanford University, 1987) and metallurgical engineering (B.S. Michigan Technological University, 1985). Since joining Cal Poly in 1991, she has received Cal Poly's Distinguish Teaching Award (2002-2003), the Northrop-Gumman Excellence in Teaching and Applied Research Award (2000-2001), the American Society for Engineering Education Dow Outstanding New Faculty Award (1997), and the TRW Excellence in Teaching Award (1992-1993).

John Warner received his B.S. in chemistry from the University of Massachusetts, Boston, his M.S. and Ph.D. from Princeton in organic chemistry. He worked at the Polaroid Corporation for nine years, and then went to UMASS Boston, where he has started the world's first green chemistry Ph.D. program. He is now at the University of Massachusetts, Lowell, where he directs a large research group working on a diverse set of projects involving green chemistry, using principles of crystal engineering, molecular recognition, and self-assembly. His work combines aspects of community outreach, government policy, and industrial collaboration. He is associate editor of the journal *Organic Preparations and Procedures International* and on the editorial board of *Crystal Engineering* and *Crystal Growth and Design.* He recently received the 2004 Presidential Award for Excellence in Science Mentoring from President Bush and the Outstanding Service to Nursing Award from Sigma Theta Tau International Honor Society of Nursing. He was awarded the American Institute of Chemistry's Northeast Division's Distinguished Chemist of the Year for 2002. His recent patents in the fields of semiconductor design, biodegradable plastics, personal care products, and polymeric photoresists are examples of how green chemistry principles can be immediately incorporated into commercially relevant applications. Professor Warner is coauthor of the book "Green Chemistry: Theory and Practice" and serves on the Board of Directors of the Green Chemistry Institute in Washington, DC.

Appendix E

Workshop Attendees

Spiros N. Agathos, University of Louvain, Belgium
David Allen, University of Texas, Austin, TX
Paul Anastas, Green Chemistry Institute, Washington, DC
John Andraos, York University, Toronto, ON, Canada
Eric J. Beckman, University of Pittsburgh, Pittsburgh, PA
Michael R. Berman, Air Force Office of Scientific Research, Arlington, VA
James Brown, American Chemical Society, Washington, DC
Edward J. Brush, Bridgewater State College, Bridgewater, MA
Kevin Carroll, House Science Committee, Washington, DC
Michael C. Cann, University of Scranton, Scranton, PA
Amy S. Cannon, Rohm & Haas Electronic Materials, Marlborough, MA
Dennis Chamot, National Research Council, Washington, DC
Terrence ("Terry") J. Collins, Carnegie Mellon University, Pittsburgh, PA
F. Fleming Crim, University of Wisconsin-Madison, Madison, WI
Berkeley ("Buzz") W. Cue, Jr., Consultant (formerly of Pfizer), Ledyard, CT
Cliff I. Davidson, Carnegie Mellon University, Pittsburgh, PA
Kenneth M. Doxsee, University of Oregon, Eugene, OR
Arthur B. Ellis, National Science Foundation, Arlington, VA
Richard Engler, U.S. Environmental Protection Agency, Washington, DC
Joseph M. Fortunak, Howard University, Washington, DC
Thomas E. Goodwin, Hendrix College, Conway, AR
Michael H. Gregg, Virginia Polytechnic Institute and State University, Blacksburg, VA
Liz Gron, Hendrix College, Conway, AR
Jeff Gunnulfsen, SOCMA, Washington, DC
Julie Haack, University of Oregon, Eugene, OR
Sharon L. Haynie, DuPont Company, Wilmington, DE
Alan Hecht, U.S. Environmental Protection Agency, Washington, DC
Miriam Heller, National Science Foundation, Arlington, VA
Susan H. Hixson, National Science Foundation, Arlington, VA
Patricia A. Hogan, Suffolk University, Boston, MA
Rawle I. Hollingsworth, Michigan State University, East Lansing, MI
Steven M Howdle, University of Nottingham, Nottingham, England
James E. Hutchison, University of Oregon, Eugene, OR
Roshan Jachuck, Clarkson University, Potsdam, NY
Margaret E. Kerr, Worcester State College, Worcester, MA
John Leazer, Merck & Co., Rahway, NJ
Irvin J. Levy, Gordon College, Wenham, MA
Stephen Lingle, U.S. Environmental Protection Agency, Washington, DC
Stanley E. Manahan, University of Missouri, Columbia, MO
Lon J. Mathias, University of Southern Mississippi, Hattiesburg, MS
Sean McGinnis, Virginia Tech University, Blacksburg, VA
Tyler McQuade, Cornell University, Ithaca, NY
James R. Mihelcic, Michigan Technological University, Houghton, MI
Ty Mitchell, National Science Foundation, Arlington, VA
Karen Peabody O'Brien, Green Chemistry Institute, Washington, DC
Kathryn Parent, Green Chemistry Institute, Washington, DC
Kathleen Parson, National Science Foundation, Arlington VA

Alvise Perosa, Università Ca' Foscari, Venice, Italy
James Rea, Green Chemistry Institute, Washington, DC
Laura Ruth, Consultant, Worcester, MA
John C. Saddler, Pfizer Inc., Kalamazoo, MI
Darlene S. Schuster, AIChE Institute for Sustainability, New York, NY
David R. Shonnard, Michigan Technological University, Houghton, MI
Greg Smith, SOCMA,Washington, DC
Jim Solyst, American Chemistry Council, Arlington, VA
Gary O. Spessard, St. Olaf College, Northfield, MN
Martin ("Marty") Spitzer, House Science Committee, Washington, DC
Jeffrey I. Steinfeld, Massachusetts Institute of Technology, Cambridge, MA
Eileen Stephens, Green Chemistry Institute, Washington, DC
Bala Subramaniam, Center for Environmentally Beneficial Catalysis, Lawrence, KS
Xiuzhi Susan Sun, Kansas State University, Manhattan, KS
Richard M. Taber, National Academy of Engineering, Washington, DC
Dicksen Tanzil, BRIDGES to Sustainability, Houston, TX
David Robert Taschler, Air Products & Chemicals Inc., Allentown, PA
Javad Tavakoli, Lafayette College, Easton, PA
Linda Vanasupa, California Polytechnic State University, San Luis Obispo, California
Jorge Vanegas, Texas A&M University, College Station, TX
John Warner, University of Massachusetts, Lowell, MA
Denyce Wicht, Suffolk University, Boston, MA
Michael P. Wilson, University of California, Berkeley, Berkeley, CA
James Wishart, Brookhaven National Laboratory, Upton, NY
Frankie Wood-Black, Conoco-Phillips, Houston, TX
Richard P. Wool, University of Delaware, Newark DE
Jennifer Young, Green Chemistry Institute, Washington, DC
Julie Beth Zimmerman, U.S. Environmental Protection Agency, Washington, DC

Appendix F

Origin of and Information on the Chemical Sciences Roundtable

In April 1994 the American Chemical Society (ACS) held an Interactive Presidential Colloquium entitled "Shaping the Future: The Chemical Research Environment in the Next Century."[1] The report from this colloquium identified several objectives, including the need to ensure communication on key issues among government, industry, and university representatives. The rapidly changing environment in the United States for science and technology has created a number of stresses on the chemical enterprise. The stresses are particularly important with regard to the chemical industry, which is a major segment of U.S. industry, makes a strong, positive contribution to the U.S. balance of trade, and provides major employment opportunities for a technical workforce. A neutral and credible forum for communication among all segments of the enterprise could enhance the future well-being of chemical science and technology.

After the report was issued, a formal request for such a roundtable activity was transmitted to Dr. Bruce M. Alberts, chairman of the National Research Council (NRC), by the Federal Interagency Chemistry Representatives, an informal organization of representatives from the various federal agencies that support chemical research. As part of the NRC, the Board on Chemical Sciences and Technology (BCST) can provide an intellectual focus on issues and fundamentals of science and technology across the broad fields of chemistry and chemical engineering. In the winter of 1996, Dr. Alberts asked BCST to establish the Chemical Sciences Roundtable to provide a mechanism for initiating and maintaining the dialogue envisioned in the ACS report.

The mission of the Chemical Sciences Roundtable is to provide a science-oriented, apolitical forum to enhance understanding of the critical issues in chemical science and technology affecting the government, industrial, and academic sectors. To support this mission the Chemical Sciences Roundtable will do the following:

- Identify topics of importance to the chemical science and technology community by holding periodic discussions and presentations, and gathering input from the broadest possible set of constituencies involved in chemical science and technology.
- Organize workshops and symposiums and publish reports on topics important to the continuing health and advancement of chemical science and technology.
- Disseminate information and knowledge gained in the workshops and reports to the chemical science and technology community through discussions with, presentations to, and engagement of other forums and organizations.
- Bring topics deserving further in-depth study to the attention of the NRC's Board on Chemical Sciences and Technology. The roundtable itself will not attempt to resolve the issues and problems that it identifies—it will make no recommendations, nor provide any specific guidance. Rather, the goal of the roundtable is to ensure a full and meaningful discussion of the identified topics so that the participants in the workshops and the community as a whole can determine the best courses of action.

[1]American Chemical Society. *Shaping the Future: The Chemical Research Environment in the Next Century*. American Chemical Society Report from the Interactive Presidential Colloquium, April 7-9, 1994, Washington, DC.